Religion
and
Global Climate Change

Religion and Global Climate Change

A Handbook for Faith Leaders and Climate Activists

Frederick W. Krueger

editor

January 7, 2015

The Symbolic Meanings
of the Front and Back Cover

THE COVER PHOTO DEPICTS A CITY SWELTERING UNDER A MOISTURE LADEN HAZE. This reflects the higher humidity that people will experience in a warmer world. If we do not curb growing levels of carbon dioxide, the average world temperature will rise and cause increased humidity in some areas and greater aridity in others. Those areas which experience more moisture will see more flooding, more torrential rain, and more intense storms. Those in arid areas will experience an increase in wild fires such as are already occurring in the western United States, plus expanding desertification, more frequent drought, and lower agricultural yields.

The vertical strip down the left margin is an ancient artistic design which reminds viewers of the traditional religious vision of the world. The points of light, rather like sparks, represent the presence of God and the Holy Spirit filling the world. This particular design is copied from the ceiling of the tomb of Gallus Placidius, granddaughter of Roman Emperor Constantine. The original design is located outside the Church of San Vitaly in Ravenna, Italy, and dates from the 4th century. This symbol was a reminder that we all live in the presence of God and serve our Creator as we care for the world and its integrity.

The back cover shows a map of the world with "WARNING" emblazoned across it. This reminds us that our earth is in serious danger. We are burning fossil fuels and contaminating the earth's atmosphere with measurably increasing levels of heat-trapping greenhouse gases. Unless we cease our profligate use of high energy fuels and find ways to embrace clean sources of energy, we may see those letters change from "Warning" to "NO LONGER FIT FOR HUMAN HABITATION." The choice is ours. Let us change our ways now while there is still time to change.

The National Religious Coalition on Creation Care

1100 Hughes Avenue • Santa Rosa, California 95407

www.NRCCC.org

Contents

Foreword   **Why this Handbook?**   vii

Section 1   **The Faces of Global Climate Change**   9

          Photographic Images of a Growing Crisis

Section 2   **Religious Declarations on Climate Change**   35

          Formal Statements from Faith Organizations

Section 3   **Faith Leaders Speak Out**   89

          The Need to Address Climate Change

Section 4   **Why the Urgency on Climate Change?**   131

          Scientific and Other Perspectives

Section 5   **What You and Your Congregation Can Do**   185

          Recommendations for Action

Foreword

THIS HANDBOOK IS DEDICATED TO RELIGIOUS LEADERS, CLERGY, AND climate activists everywhere.

Religious leaders, across America and around the world, are becoming a unified voice on the urgent need to address global climate change. This is happening because top religious leaders have studied this issue. They realize that we must change our ways and stop the rise of carbon dioxide that is causing climate change. But at the local level, the insights of religious leadership are not well heard. This book assembles the major religious voices on climate change and presents their formal declarations in the hope that we can awaken those who have not yet studied the critical dimensions of our climate predicament. It is further hoped that by including sound science, we can bridge the perspectives between faith institutions, scientific organizations, and civil society.

This anthology presents the basic religious understanding and commentary on climate change. This will build conviction and resolve to address our climate predicament. This should aid in developing local and regional coalitions to enable religious leaders, scientists, medical professionals and activists to work together and to form vigorous coalitions to engage the global evil of climate change.

The stakes are high. Civil society is facing a historic crossroads. We need to reach across social separations and teach the urgency of our predicament. You can help – by telling those you know about our climate plight, and by embracing the lifestyle changes that reduce your carbon footprint. We are addressing the problem of a world out of balance and a humanity still unaware of the challenge of turning society so that it can correct the destructive course on which we are headed.

The choice ultimately comes down to what William Shakespeare wrote in the words of his character Hamlet in the play Macbeth:

> To be, or not to be, that is the question:
> Whether 'tis nobler in the mind to suffer
> The slings and arrows of outrageous fortune,
> Or to take arms against a sea of troubles....

The Faces of Global Climate Change

OUR WORLD IS RAPIDLY CHANGING. As greenhouse gases increase in the atmosphere, new conditions are emerging that impact people and the natural world. Examine the following photos and read the brief captions. These images depict the warning signs that tell us climate change is already here. They tell us that our way of living, particularly how we have put society together, is causing harm to people and the land on which we depend for livelihood. Human society is now on a collision course with disaster. Just as a person driving down the road who sees an obstacle ahead will slow down and stop, so these images tell us that carbon pollution of the atmosphere and greenhouse gases are causing great harm and that we have to slow down and stop – now.

The images on the next pages reveal the breadth of the impacts taking place. The range of issues is as broad as the world. People are already suffering and dying prematurely. Land, air, water, forests, and food are under assault from climate changes; so are cities and communities worldwide. Public health, storm intensities, water supplies, and an increasing heat and drought are all involved. What sort of a world are we leaving to our children? This is just the beginning of the sorrows that climate change is bringing to the world.

As you view these pictures, know that what is happening now in distant corners of the globe will soon be at our doorsteps. This asks the question of how do we deal with this? What is our responsibility – to our families, to our neighbors and to the future of our children? What does God call of us in this situation? What are the questions that we should be asking ourselves?

This first section introduces the initial wave of issues and provides details about the climate disruptions that are already taking place now. This will reveal the depth of the climate problem and show why we must galvanize ourselves and society to do all that we can to slow and stop the progression of forces that are creating this historically unprecedented situation.

Earth's Atmosphere is Changing

The atmosphere of Earth is a thin layer of less than twenty miles above the surface of our planet. This can be seen in the thin white band above the earth and below the blue of space in this photo taken from the Discovery Space Shuttle while approaching the Japanese Island of Hokkaido at dusk. Carbon dioxide (CO_2) levels are steadily increasing in the atmosphere.

Before the industrial revolution the concentration of atmospheric CO_2 was about 270 parts per million (PPM). By 1900 this had increased to 280 ppm. In 2014 this concentration has increased to 402 ppm and the rate of increase is climbing annually.

The increase of CO_2 causes more of the sun's warmth to be retained by the atmosphere. This is changing the weather and causing other changes to life on earth. The burning of fossil fuels such as oil, gas and coal are the primary culprits in the rise of atmospheric CO_2. Scientists agree the main cause of the current warming of the planet is human expansion of the "greenhouse effect" – warming that results when the atmosphere traps heat radiating from Earth toward space.

Heat Waves are Intensifying

One hot day, or even an intense heat wave, cannot be conclusively blamed upon climate change. However, the steady increase in the frequency and intensity of heat waves and the weather changes that are now taking place can be blamed on climate change.

In 2010 extremely hot weather and drought threatened lives and increased the death rate across Europe. In 2011 Texas and the U.S. Southwest experienced record breaking heat. Everywhere excessively hot days are on the increase. The combination of high humidity plus high night-time temperature can make for a deadly pairing, posing a particular threat for the elderly. Extreme heat events are responsible for heightened levels of early deaths.

The number of areas affected by extreme heat is now on the rise and the hottest temperatures on earth are increasing. Higher temperatures reduce crop yields and drive up energy usage. Heat waves are the No. 1 weather-related killer in the U.S. Evidence suggests that heat waves are becoming more likely due to climate change. Study of heat waves in Russia, Texas, and Australia show that climate change played a role in making these events more likely.

Glaciers are Melting

Dawn strikes the mountains above St. Mary's Lake in Montana's Glacier National Park. When the park was created in 1910, it had more than 150 glaciers. Now in 2014 only a few glaciers remain, and those are all significantly reduced in size. Within another twenty years, all of them will be gone.

On Mount Hood near Portland, Oregon, photos taken in late August, 17 years apart, show the rapid decline of its glaciers. Around the world glaciers store about 70% of the world's fresh water. Glaciers on every continent are now in rapid retreat.

Now things that normally happen in geologic time are happening during the span of a human lifetime.

1985

2002

© GARY BRAASCH

The Arctic is Warming

A Native Inuit woman from Baffin Island in Nunavut, Canada worries about the future. Temperatures in the Arctic region are rising much faster than those in more moderate climates. The Inuits and other Arctic people are disproportionately experiencing the effects of global warming, many of which threaten the lifestyles of these people.

Caribou, a staple of Inuit diet, are falling through the thinning ice. Hunting, whaling, and fishing are becoming more difficult as migration patterns change. Coastal erosion has increased as melting sea ice is unable to provide the protective barrier it did in the past. Rivers that are usually ice-filled are melting early and flooding low-lying areas.

The United Nations' Intergovernmental Panel on Climate Change predicts that global temperatures will rise by an additional 3° to 10° F. before the end of this century. Rising temperatures have a dramatic impact on Arctic ice, which serves as a kind of "air conditioner" at the top of the world. Since 1978 the area of Arctic sea ice has been shrinking by 9% per decade.

Coral Reefs are Dying

This is a close up photo of a bleached coral. The bright white area is the bleached portion on this otherwise healthy coral reef. Rising ocean water temperatures, one of the many effects of global warming, are threatening coral reefs around the world.

In response to warming ocean temperatures, a process known as coral bleaching occurs. Although bleaching does not immediately kill coral reefs, they quickly become more vulnerable to diseases. Warming oceans will result in decreasing levels of phytoplankton, a natural absorber of carbon dioxide. About one-quarter of all sea life depends on plankton as the foundation of the ocean food chain.

Around the world scientists estimate that 20% of the world's coral reefs have already been destroyed because of climate change. These reefs are among the most biologically diverse ecosystems on the planet. Coral reefs are the birth place of about one-quarter of all sea life. Worldwide coral reefs provide seafood for over one billion of the world's people and income valued at over $375 billion per year.

Sea Levels are Rising

This aerial photo shows the city of Male, the capital city of the Maldive Islands in the Indian Ocean.

Worldwide many heavily populated areas, including Florida in the U.S., are looking to their coastlines with growing concern as sea levels rise. But perhaps none have taken it so seriously as Male. With a maximum elevation of only 8 feet, any rise in sea level will engulf currently inhabited land. In response to this impending threat, the city has built a great seawall around the capital.

The ocean level has already increased by one foot over the last century. As climate change accelerates, the oceans will warm and continue to rise. A three-foot rise in sea level will cause large areas of the State of Florida to disappear (see the dark areas on the map). Storms will reach much further inland and have greater impact on cities and coastal areas.

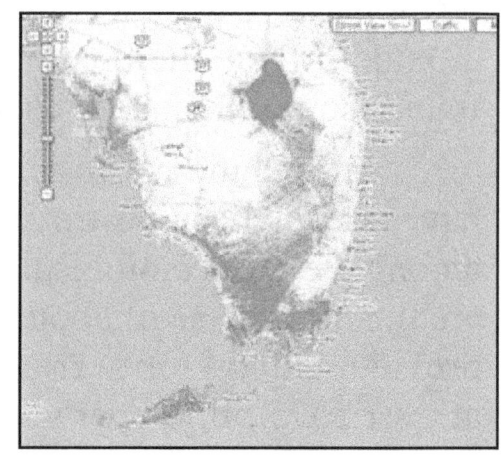

The Oceans are Becoming Acidic

As carbon dioxide levels rise in the atmosphere, about one-third of the CO^2 is absorbed into the oceans. Carbon dioxide, when mixed with sea water, produces carbonic acid. This is why ocean water is becoming more acidic.

The corals in the photo above are particularly vulnerable to carbonic acid. So are clams, oysters and most shellfish; crabs and lobsters; plankton, and some fish species. All are harmed by higher ocean acidity. The danger is that the corrosive effect of acidic oceans will trigger a dramatic shift in coastal species and jeopardize shellfish stocks, some sport fishing, and perhaps major sections of the ocean food chain.

Scientists are calling for drastic measures to avert massive bleaching of the world's coral reefs. Dr. Jane Lubchenco, a marine scientists and past director of the National Oceanic and Atmospheric Administration, has called ocean acidification global warming's "equally evil twin."

The only way to avert an ocean disaster is to limit the rise of carbon dioxide in the world's atmosphere. This means society must shift from dirty polluting fossil fuels to clean renewable sources of energy.

Droughts Are Becoming Longer

Billions of people will be at increased risk from droughts in the years ahead. The expected climatic changes in this century will intensify the hydrological cycle with rainy seasons becoming shorter and more intense in some regions, while droughts will grow longer in other areas. This will endanger food and agriculture. Starvation will be more common.

"Droughts are already more severe and widespread," says The World Water Council. "Up to 45% of reported deaths from natural disasters between 1992 and 2012 resulted from droughts and famines. The most vulnerable communities are impoverished peoples occupying marginal rural and urban environments."

Africa's recent unprecedented droughts signal widespread climate change. Australia too is entering long-term climate change which may cause longer and more frequent droughts. The on-going drought could leave the City of Sydney's dams dry in just two years. California and the Western U.S. will also face water shortages and some communities may run out water.

Exotic Species are Proliferating

A woman holds a warm water barracuda caught just off the Seattle waterfront. Ocean sunfish have been spotted off the Pacific coast of Oregon and Washington states, an area where they have never before been seen. Tropical lizard fish and other barracuda have been landed by fishermen inside Puget Sound. Other southern natives, including sardines and striped bass, have made the northward migration.

"From just these small little teases of El Niño... there has been major disruption," said Steve Jeffries, a marine mammal expert with the Washington State Department of Fish and Wildlife.

The surface of the world's seas has already warmed since the late 1800s. Even slight changes in temperature can alter the complex weave of the marine food web.

Worldwide, many bird, fish and other animal species are on the move, trying to find more hospitable climate conditions.

Storms are Growing More Powerful

This is one of thousands of hurricane-flooded homes along the coast of New Jersey. Tens of thousands of homes in the New York-New Jersey area were destroyed and residents displaced following Hurricane Sandy. Climate change is leading to more powerful, more devastating and more frequent hurricanes and floods.

Hurricanes have always plagued the Atlantic and Gulf Coasts, but global warming is making matters much worse. As sea levels rise, storm surges reach higher and move further inland. Even relatively minor tropical storms can now cause coastal flooding and result in evacuations of residents and damage to seashore properties.

New scientific evidence shows a clear link between the strength of hurricanes and global warming. As oceans warm, they produce more intense storms. Huge storms, such as Hurricanes Rita, Katrina and Sandy have increased during the past 35 years while smaller storms have become less common.

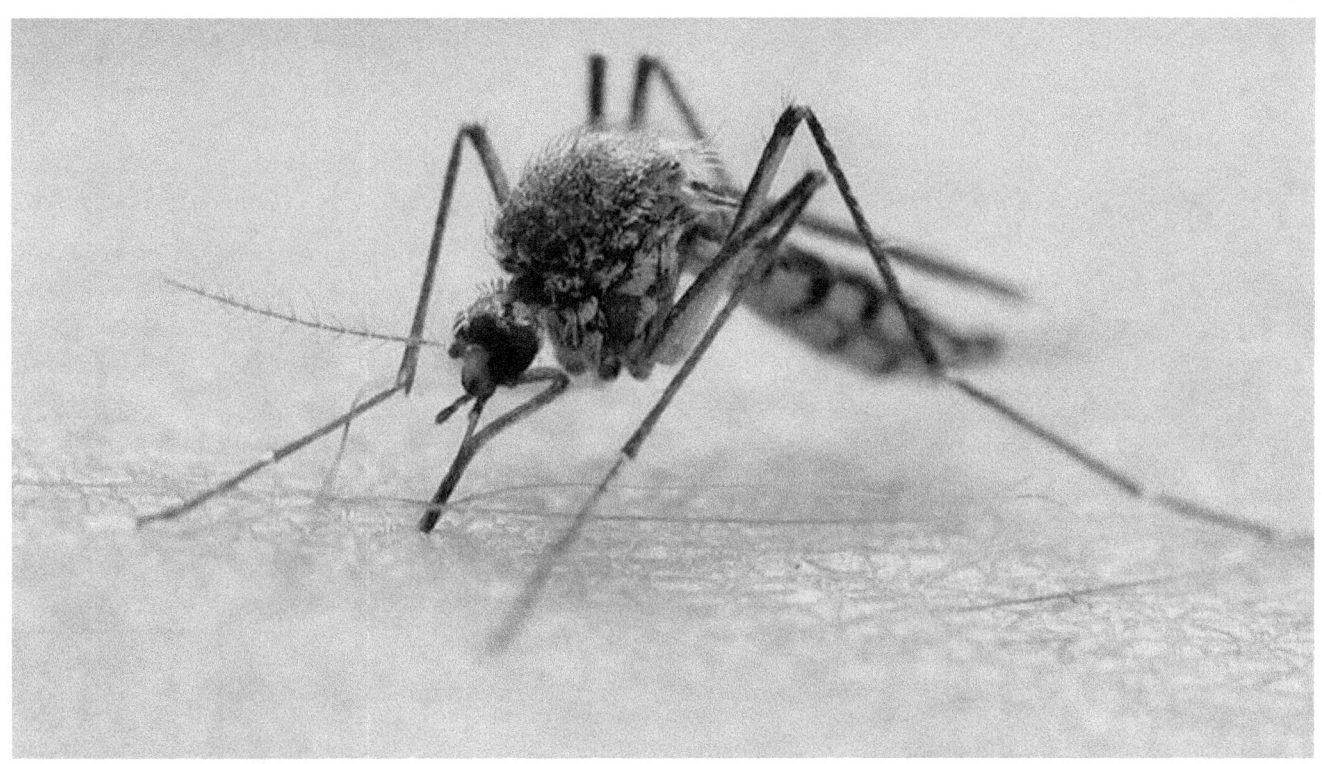

Tropical Diseases are Spreading

Global climate change – with an accompanying rise in floods, droughts and unpredictable weather – is fueling the spread of epidemics in areas unprepared for new diseases, say health experts worldwide. Mosquitoes, ticks, mice and other pathogen carriers are surviving through warmer winters and expanding their range, bringing health threats with them.

Malaria is climbing the mountains to reach populations in higher elevations in Africa and Latin America. Cholera is growing in warmer seas. Dengue fever and Lyme disease are moving north. West Nile virus, never seen on North America until 2000, has already infected more than 21,000 people in the U.S. and Canada and already killed over 800 people.

"The evidence of global warming on human health is everywhere," says Dr. Paul Epstein, MD, from Harvard Medical School. "Patterns of diseases are shifting in unexpected ways. Over thirty diseases new to medicine have emerged in just the last 30 years. This is unprecedented in the annals of medicine. This could cause overwhelming damage, especially in developing countries, with already over-stretched physical and human infrastructures."

Forests are Dying

In British Columbia and in other boreal forests pine bark beetles are spreading north as temperatures increase. They now live in places where in colder times they could not survive. The result is massive insect infestations and spreading forest death. Scientists with the Canadian Forest Service say the average winter temperature here has risen by more than 4° F. during the last twenty years.

The outbreak of bark beetle in British Columbia's forests has been one of the highest-profile forestry issues in recent years. Since 1997, bark beetle outbreaks have killed over 80 million trees stretched across more than one million acres. The damage adds to the risk of wildfire.

"It's pretty gut-wrenching," said Allan Carroll, a research scientist at the Pacific Forestry Centre in Victoria, B.C., which tracks the connection between warmer winters and the spread of the beetle. "People say climate change is something for our kids to worry about. No. It's happening now."

Catastrophic Wildfires Are Increasing

Rising temperatures are drying the forests. This has stoked an increase in large wildfires over the past 34 years as spring comes earlier, mountain snows melt sooner, and forests become tinder dry. The changing climate is the single most important factor driving a four-fold increase in the average number of large wildfires in the Western United States since 1970.

Researchers report almost seven times more forested land burned between 1987 and 2003 than during the previous 17 years. If regional temperatures continue to rise, as is predicted, wildfires will intensify. As more forests burn, the fire destruction will release massive amounts of carbon dioxide, further accelerating the increase of greenhouse gases and further increasing temperatures.

"It all fits together," says climate researcher Anthony Westerling, at the Scripps Institution of Oceanography in La Jolla, California. "The fire seasons now start earlier and run longer. This is all consistent with a warming climate." Longer droughts lead to more forest fires—as we have seen across the Western United States this past year.

The Death Rate is Rising from Heat Waves

The U.S. Center for Disease Control (CDC) finds that an average of 660 people die nationwide from heat waves each year, making it the leading cause of weather-related mortality in the U.S. The CDC defines heat waves as "several days of temperatures greater than 90° F; warm, stagnant air masses; and consecutive nights with higher-than-usual minimum temperatures." A severe heat wave in Chicago was blamed for 700 deaths, and perhaps as many as 333 people died in California in July 2006 as the state was gripped in unrelenting heat.

Researchers from the Rollins School of Public Health at Emory University in Atlanta have been using the latest temperature projections to understand the rising human toll of a warmer future.

They find that as heat waves increase in length and severity, high temperatures will kill ten times more people than they did back in 2000. In 2002-2004, an average of 187 people in the eastern third of the U.S. succumbed to heat waves. Within several decades, that number will increase ten fold to well over 6,000 per year.

In Australia, heat stress is already the number one natural killer, accounting for more human deaths than floods, cyclones, bushfires and storms combined. It's a situation which will only become worse as the climate changes.

Traditional Life Styles Will Change

In Barrow, Alaska, the northernmost U.S. town 350 miles above the Arctic Circle, the 1,800 residents of this Eskimo village do not see the sun from late November until January. The Arctic is now warming at a rate much faster than the rest of the planet. Average temperatures in Barrow have increased by over 4°F during the last 30 years. In western Canada and eastern Russia, temperatures are up from 4 F. to 7 F. over the last 50 years, a rate more than double the global average. The warming temperatures in Siberia will dry out peat bogs, which are natural carbon sinks that can help to offset fossil fuel emissions.

Scientists have found that the ice in the Arctic Ocean is melting so rapidly that most of it could be gone within twenty years. The results could be catastrophic: Low-lying lands would be inundated by rising sea levels. If present trends continue, the Arctic will be ice-free by 2030. The impacts of dwindling ice cover in the Arctic are far-reaching, from species endangerment to enhanced global warming, to the weakening or shut-down of global ocean circulation.

Animal Species Will Become Extinct

A polar bear walks across bare rocky ground near Wager Bay, Canada. Perhaps the Arctic's most charismatic large mammal, polar bears face serious threats from global warming. The bears depend on sea ice as a platform from which they can hunt seals, their main prey. As more sea ice melts and polar bears are left with rocky ground like that shown in this photograph, hunting for food becomes increasingly difficult for these large mammals. Many are already dying from food shortages; others are drowning as the sea ice melts.

Like many other species, including the great white shark, polar bears are now moving northward as temperatures increase. Whether these species can outrun warming temperatures is unclear, as some scientists say that global warming will lead to the extinction of millions of species, including polar bears and many birds and amphibians over the next 50 years. Worldwide biologists predict that over 35% of all plant and animal species in the world will likely perish.

World Wheat Crop is Threatened by Global Warming

Extreme temperatures are cutting into India's wheat yields. It will be much more difficult to feed everyone in a warmer world. Two-thirds of wheat in poor countries, and 23 percent in rich countries – nearly half of the entire world's total crop – is now at risk from global warming.

Stanford University researchers warn that current projections under-estimate the extent to which hotter weather in the future will accelerate this process. This underscores the challenge of feeding a rapidly growing population as the world continues to warm.

Across India's breadbasket on the Ganges plain, winter wheat is planted in November and harvested as temperatures rise in spring. David Lobell of Stanford University used nine years of images from the MODIS Earth-observation satellite to track when wheat turned from green to brown, a sign that the grain is no longer growing. He found that the wheat turned brown earlier when average temperatures were higher, with spells over 34 °C having a particularly strong effect.

"This is an early indication that a situation that is already bad could become even worse," says Andy Challinor of the University of Leeds, UK. Yet global yields need to rise 50 percent to feed India's growing population.

Climate Refugees will Increase

Whether the cause is drought and food shortages, rising seas or flooding from torrential rains, climate change will dramatically increase the number of refugees from their ancestral homelands.

Since 1990 the Intergovernmental Panel on Climate Change (IPCC) has said that the single greatest impact of climate change might be human migration — with millions of people forced into relocation by shoreline erosion, coastal flooding and agricultural disruption. Since that time, reports show that climate change is poised to become a major driver of global population displacement — a monstrous crisis in the making.

U.N. officials estimate that a third of Bangladesh's coastline could be flooded if the oceans rise by one meter, something projections anticipate. This will create upwards of 20 million displaced Bangladeshis.

Southern Florida will experience a similar requirement for relocation as storm surges will inundate coastal areas. Boston, New Orleans, New York and other coastal areas are also in jeopardy.

Warmer Water Disrupts Pacific Food Chain

On the Pacific Ocean just off San Francisco, a sailboat studies the decline of a bird species on offshore islands. This has rekindled fears that global warming is undermining coastal food supplies.

Tiny Cassin's auklets live much of their lives on the open ocean. But each spring, these relatives of the puffin venture to isolated outposts like the Farallone Islands to lay their eggs. Adult auklets feed their chicks with krill, the tiny crustaceans that anchor the ocean's food web. But not this year. Almost none of the 20,000 pairs of nesting auklets here will raise a chick in 2012 that lives for more than a few days.

Scientists blame changes in the climate. The seas are warmer. The number of krill being produced is lower. "One of the things that the climate models predict is unpredictable weather, extreme weather, and that the whole seasonal cycle of events will not be what we expect," said Bill Peterson, a NOAA oceanographer. The absence of krill has also led to a collapse of the juvenile rockfish population. The auklet is unlikely to adapt to the sudden loss of its main food source. Other animals could follow. "In the worst case scenario," Peterson declared, "we could see a great depression of the entire ecosystem."

Insect Explosions will Threaten Food Crops

Locusts swarm over a field in the Canary Islands. Biologists have discovered that as a result of rising levels of carbon dioxide in the atmosphere, the numbers of leaf-eating insects will surge at a time when crop production will have to be boosted to feed an extra three billion people living at the end of 21st century.

The anticipated 5° C rise in global temperatures caused by a doubling of atmospheric CO_2 levels will send insect numbers soaring and cause potentially major disruptions to agriculture.

Dr. Ellen Currano of Pennsylvania State University, lead author of the study, said, "Our study shows [that] … when temperature increases, the diversity of insect-feeding damage on plant species also increases." This means food supplies will likely shrink precisely at a time when worldwide populations are growing.

A Sea Lane is Opening over the Arctic Ocean

As global temperatures rise, the impact is greater in some regions than in others. Nowhere is the impact more apparent than in the Arctic. In this photo the once impassible sea ice has melted so much that German ships have pioneered a Northeast Passage from Korea over the Arctic to Europe.

The good news for ship owners, is that the Arctic passage slashes time and money for mariners. This development could cause an economic boom for Russia. Ship owners say they plan to use this new route on a regular basis as it saves ten days and $3 million over the normal 11,000 nautical-mile voyage from Korea through the Indian Ocean to the North Atlantic.

The bad news is this feat is only possible because the Arctic icecap is melting at an alarming rate, leaving vast stretches of open water where solid pack ice frustrated earlier attempts at navigation. Since 2009 thru 2014, each year has seen lower levels of Arctic sea ice. This is what allows the recently initiated navigation of the Arctic.

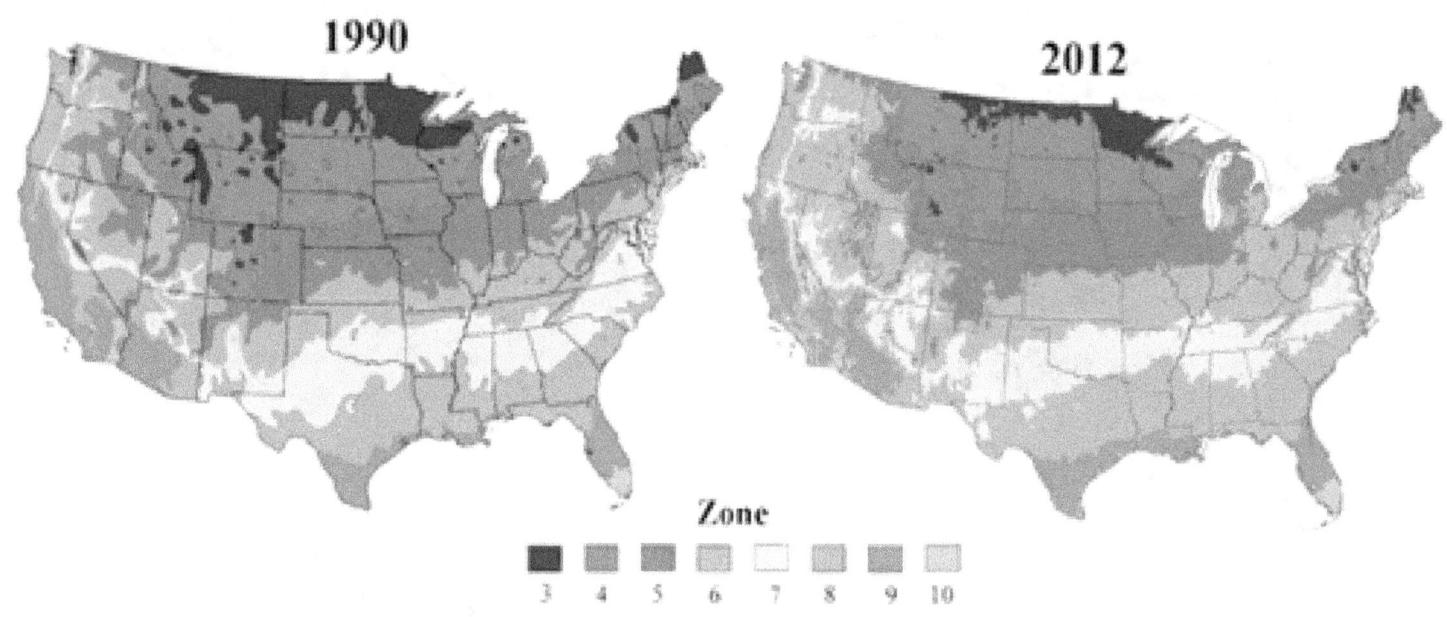

Image courtesy U.S. Department of Agriculture

USDA Plant Hardiness Zone Map Changes

In recognition of warmer temperatures and longer growing seasons, the U.S. Department of Agriculture has shifted its plant growing hardiness zone boundaries by an average of one 5-degree Fahrenheit half-zone warmer throughout most of the U.S. That shift may seem like a small thing, but it's a huge difference for everyone from home gardeners to big agriculture.

On the plus side, it means certain crops can now be safely grown in places farther north than ever before.

On the downside, it means that certain other crops can't be grown as far south as they used to be. More specifically this means that large swaths of presently borderline arable land in the U.S. Southwest will become useless-for-agriculture desert. This also means expensive disruptions to the agricultural status quo. Farmers in this borderland will be displaced. Grain production will decline. Food prices will rise.

As climate change accelerates, this shift will not be the last time these hardiness zones need to be updated to reflect contemporary reality.

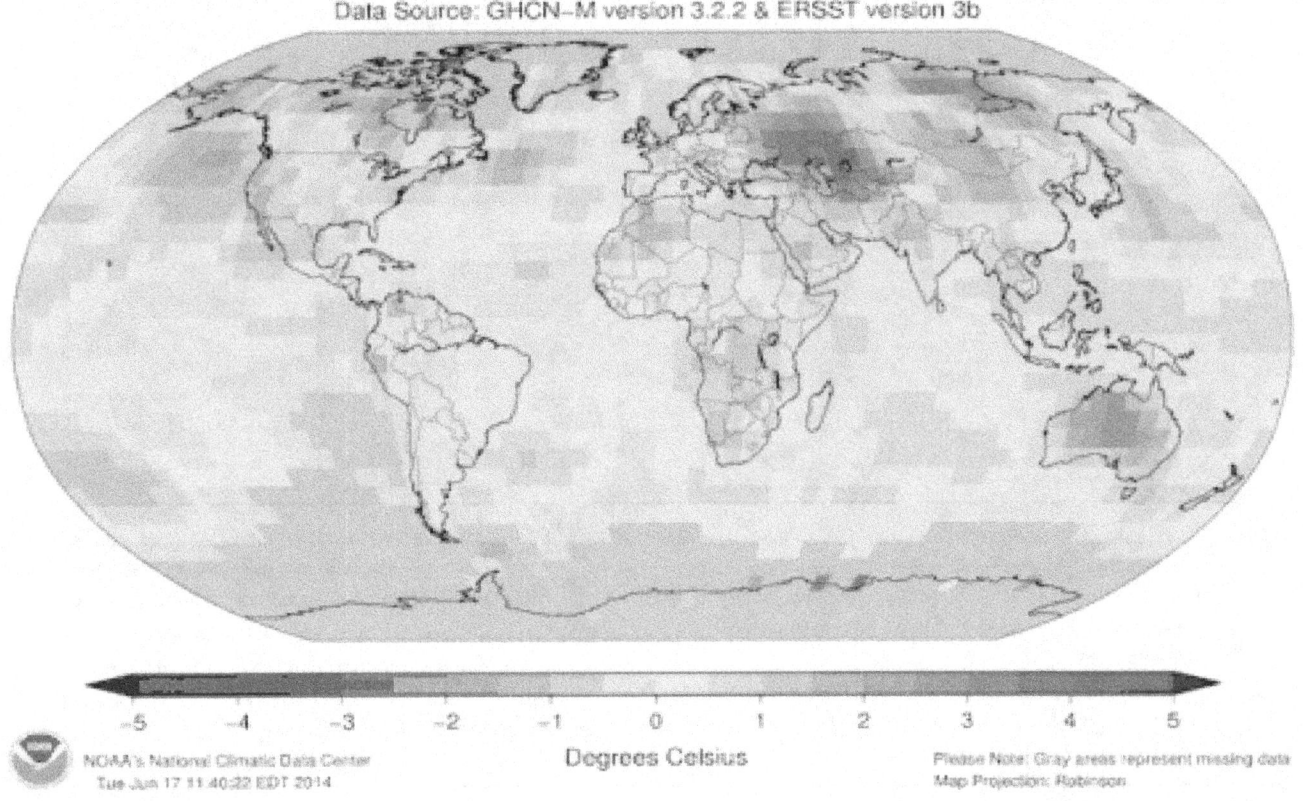

Land & Ocean Temperature Departure from Average May 2014
(with respect to a 1981–2010 base period)
Data Source: GHCN-M version 3.2.2 & ERSST version 3b

NOAA's National Climatic Data Center
Tue Jun 17 11:40:22 EDT 2014

Degrees Celsius

Please Note: Gray areas represent missing data
Map Projection: Robinson

2014 On Track To Be Hottest Year on Record

The above graph from National Climate Data Center of the National Oceanic and Atmospheric Administration (NOAA) has just announced that the year 2014 has so far been the hottest year on record.

Even though the year is not yet over, a series of warm months — including the warmest May, June, August and September on record, plus an April that was the second hottest month on record – have set one question circulating: Will 2014 take the title as the warmest year on record?

"2014 could become the warmest on record, depending on how things play out," said Kevin Trenberth, a climate scientist with the National Center for Atmospheric Research in Boulder, Colorado. For 357 months in a row, global temperatures have been warmer than average. The record-setting May came in with an average temperature 1.33°F above the 20th century average, according to figures from the National Oceanic and Atmospheric Administration. The implication of these measurements is that the impacts of climate change are accelerating rapidly.

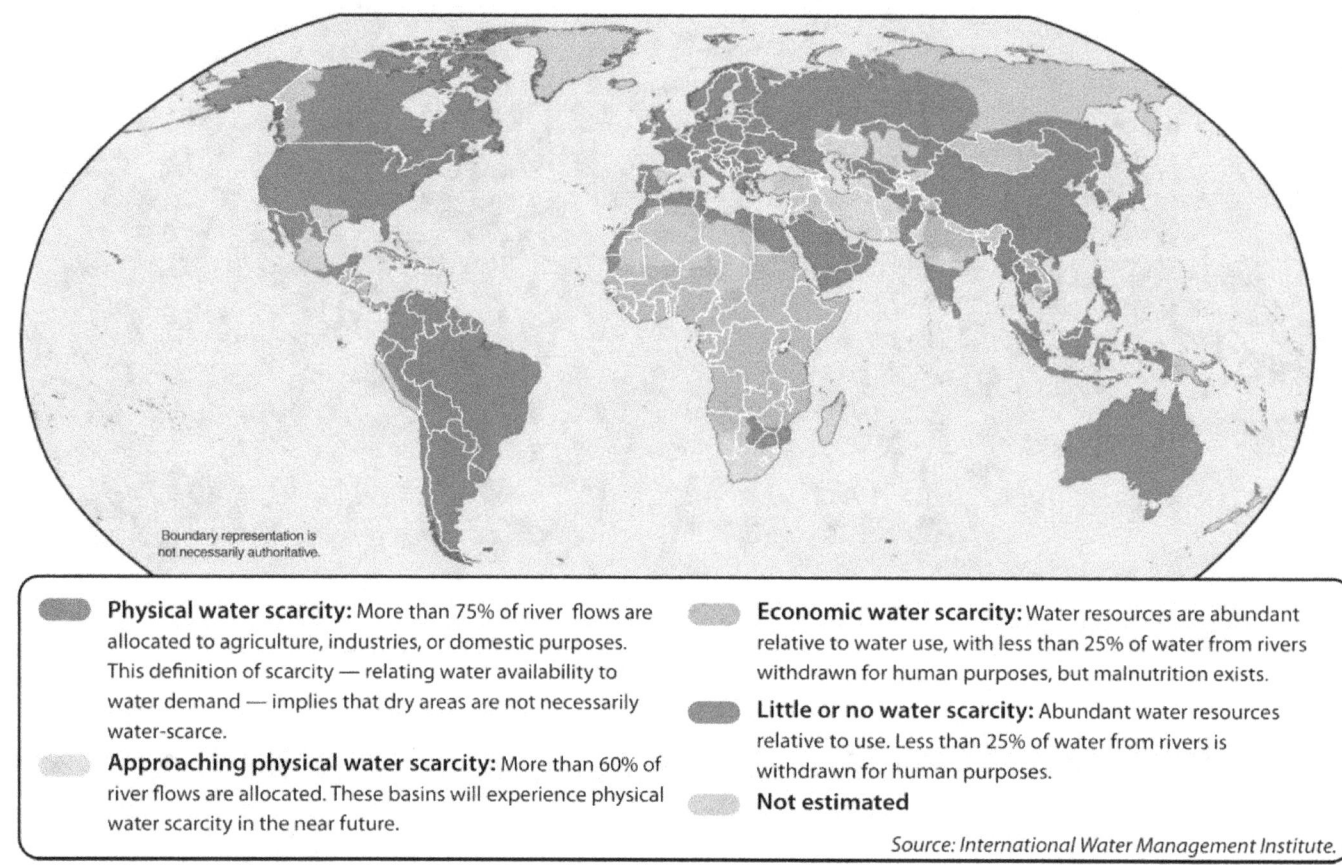

Boundary representation is not necessarily authoritative.

Physical water scarcity: More than 75% of river flows are allocated to agriculture, industries, or domestic purposes. This definition of scarcity — relating water availability to water demand — implies that dry areas are not necessarily water-scarce.

Approaching physical water scarcity: More than 60% of river flows are allocated. These basins will experience physical water scarcity in the near future.

Economic water scarcity: Water resources are abundant relative to water use, with less than 25% of water from rivers withdrawn for human purposes, but malnutrition exists.

Little or no water scarcity: Abundant water resources relative to use. Less than 25% of water from rivers is withdrawn for human purposes.

Not estimated

Source: International Water Management Institute.

Water Supplies Will Shrink

Conflicts over supplies of water will increase over the next ten years, especially across the Western United States. The map above shows regions where water supply conflicts are likely to occur based on population trends, agriculture, the needs of cities and endangered species' requirements. The darker areas indicate where conflicts are most likely to occur. This analysis does not factor in future effects of climate change, which is expected to exacerbate many of these already-identified issues.

In the U.S. areas with the highest population growth in the Southwest will coincide with at risk water supplies caused by climate change.

Decreased agricultural output caused by global warming is projected. Climate change will exacerbate resource scarcities. The impact of climate change on water will vary by region. Regional differences in agricultural production are likely to become more pronounced over time with declines disproportionately concentrated in developing countries, particularly those in Sub-Saharan Africa and South Asia.

Religious Declarations on Global Climate Change

ORGANIZED RELIGION HAS STUDIED THE PROBLEM OF CLIMATE CHANGE in great depth and detail. The consistent conclusion from scores of religious consultations, panels, examinations, committee investigations and hearings is that the science is accurate and true; that the problem is urgent; that moral and ethical principles are necessary to engage the enormity and urgency of the situation; and that every person has a role and a responsibility in this steadily progressing crisis. We will all be affected – there are no exceptions.

The statements which follow are all different. Some emphasize the seriousness of the problem; others focus on the theology which causes religious groups to address this predicament; others prioritize direction and remedial action to slow the progress of climate change. All acknowledge that this is a human problem, created by the overuse and abuse of fossil fuels. The transition to clean sources of energy is important so that we can stop the measurable changes now taking place in the planet's atmosphere.

Religious institutions have made great effort to ensure scientific accuracy and spiritual authenticity in these statements. Read them to absorb the wisdom and insight that infuses each institution's effort to bring the highest standards of integrity and credibility to their conclusions.

The result is a unified religious voice that declares the urgency of global climate change. This united voice concludes that we have created a world condition that cannot be ignored if we care about the health of the planet into the future. A clear conclusion emerges from these religious declarations. Our present predicament requires a change in how we live. In particular we must use clean energy and eliminate those actions which are causing the earth to warm and its climate to change. Among the churches and religious groups which have studied this problem, which includes the large majority of people of faith, there is no controversy about these conclusions.

American Baptist Churches USA

Serving as the Hands and Feet of Christ

RESOLUTION
ON GLOBAL WARMING

THE PROBLEM

May, 2013

The report from the Second World Climate Conference, held in Geneva in 1990, stated: "If the increase in greenhouse gas concentrations is not limited, then climate change would place stresses on the natural and social systems unprecedented in the past 10,000 years."

It is believed that increased levels of greenhouse gases are causing the earth's atmosphere and surface to become warmer. The effect is similar to that of glass panels in a greenhouse.... Much of the increase is directly attributable to human industrial activity.

The Second World Climate Conference included over 700 scientists.... The scientists reached substantial agreement on a wide range of issues. The conference declared: "Emissions from human activities are substantially increasing atmospheric concentrations of the greenhouse gases. These increases will enhance the greenhouse effect, resulting on average in an additional warming of the earth's surface."

The major greenhouse gases and their sources are:

* Carbon dioxide (CO^2) generated as a by-product of everyday energy consumption, accounts for 55% of all greenhouse gases,

* Chlorofluorocarbons (CFC's) found in solvents, air conditioning fluids, refrigerants, and foam products constitute another 24% of total greenhouse gases,

* Methane (CH^4) from animal waste, rice and other cultivation, and leaked or flared during extraction of oil, gas and coal, amounts to 15% of greenhouse gases,

* The remainder consisting of nitrous oxide (N^2O) from nitrogen fertilization, nitrogen oxides (NOx) and tropospheric ozone from automobile exhaust, coal combustion and other sources, amounts to 6% of the total.

* The particularly dangerous role of chlorofluorocarbons (CFC's) deserves special mention. These gases are the primary cause of the depletion of the ozone layer in the stratosphere and contribute to warming. The ozone layer filters ultraviolet radiation. Its destruction leads to increased exposure and significantly increased levels of skin cancer.

The consequences of global warming include the melting of polar ice caps and the rise in sea levels. Such rises could inundate land that is densely populated and submerge island nations in the South Pacific and elsewhere. Approximately 50% of the world's population lives in coastal areas. Other impacts include increased storm intensity, changes in water condition and availability, stresses on health conditions, and disruption to agriculture and food production. In the words of the Second World Climate Conference: "...the impacts will be felt most severely in regions already under stress, mainly in developing countries."

Although the initial impact of these global warming trends will be felt by less industrialized nations, 75% of all carbon dioxide emissions come from more industrialized nations.

THEOLOGICAL CONSIDERATIONS

As American Baptist Christians, our earlier Policy Statement on Ecology reminded us of our responsibility to God for the care of creation (Genesis 1:1, 11-12) and of God's displeasure with humanity's misuse of creation. Further reflection calls us to consider more seriously the implications of God's call to "love your neighbor as yourself."

There is but one Creator. As travelers on this globe together, we are dependent upon Earth for sustenance. We must understand what it means to respect all that God has created and to be our neighbor's keepers. We need to expand our hearing of Jesus' "new commandment" to "Love one another." We must see the whole creation as our neighbor. As human beings we ... live in an environment we call air. It is in us as well as around us. When it is poisoned and polluted (Isaiah 24:5-6), we and all creatures are harmed. The Old Testament word for air is the same as "wind" and "Spirit." When we limit our understanding of God's Spirit, we limit our understanding and care for God's creatures and creation.

Therefore, based on our faith in the Creator God who makes us a part of a unified creation, the General Board of the American Baptist Churches, USA, calls on national boards, regions, American Baptist institutions, congregations and individuals to:

A. Join in ways to build a culture that can live in harmony with God's creation by:

1. Deepening our biblical understanding of creation and our role in preserving the gifts God has given through educational materials, courses, programs, and personal study.

2. Developing a spirituality that embraces the dignity of the character of creation, connecting our understanding of personal salvation with stewardship of God's creation.

5. Learning about the causes of global warming through self-education and inclusion of materials in church school and learning institutions at all levels, from nursery to university.

B. Join in global, local and personal efforts to safeguard world's atmospheric integrity by:

1. Building and renovating our homes and church facilities to be energy efficient....

3. Using public transportation, car pooling, and telephone conferencing.

4. Becoming ecologically aware consumers by using products, including food, that consume less energy in production, transportation, packaging, and use.

C. Address the causes and reverse the consequences of global warming by:

1. Advocating the passage of legislation at all levels to reduce carbon dioxide output and to set reduction targets for other greenhouse gases.

2. Supporting the passage of mandatory higher fuel efficiency for new vehicles.

3. Supporting rail and other means of increased transportation efficiency....

4. Combating deforestation domestically and internationally....

5. Sponsoring and supporting shareholder resolutions to corporations on issues like reduction of carbon dioxide and other greenhouse gases, phasing out of CFC's, increased energy efficiency and fuel conservation, and other issues affecting global warming.

6. Calling for an international treaty... on global warming with specific targets for the reduction of greenhouse gases.

Notes:

1 The Churches Role in Protecting the Earth's Atmosphere: Report of an Ecumenical Consultation of Churches in Northern Industrialized Countries held at Gwatt, Switzerland, from January 13-18, 1991, page 5.

2 According to the "Justice, Peace and Integrity of Creation" document, Seoul, Korea 1991, suggests a 3% annual reduction.

Adopted by the General Board of the American Baptist Churches - November 1991
161 For, O Against, 1 Abstention (General Board Reference # - 8189:6/91)

Church of the Brethren

Continuing the Work of Jesus. Peacefully. Simply. Together.

Climate Change and Christian Witness

THE THREAT TO CREATION BY GLOBAL WARMING/CLIMATE CHANGE is a cause for concern for everyone on the planet, but for Christians the issue is more than a matter of self preservation; it is a matter of faithfulness.

Throughout its history, the Church of the Brethren has been concerned with good stewardship of God's creation. We are increasingly aware of the interrelatedness of all life. We acknowledge that energy use is linked to the ecological crisis facing the Earth, the health consequences for us and future generations, and spiritual well-being in relation to other species and our Creator.

Climate change is an issue of justice. The industrialized nations, representing less than 20 percent of the world's population, are responsible for 75 to 80 percent of the annual greenhouse gas emissions. Yet those who live in poor and developing countries will be most seriously affected by global warming.

A major challenge facing humankind is an equitable standard of living for this and future generations: adequate food, water, energy, safe shelter and a healthy environment. Human-induced climate change, along with land degradation, loss of biological diversity, and stratospheric ozone depletion, threatens our ability to meet these basic human needs.

An overwhelming majority of scientific experts, while recognizing that scientific uncertainties exist, believe that Hunan-induced climate change is occurring. Indeed, during the last few years, many parts of the world have suffered heat waves, flood, droughts, fire, and extreme weather events leading to significant economic losses and loss of life. In the past century, much of the world's polar and mountain ice has melted, and in the past few decades the melting has accelerated. While individual events cannot be directly linked to human-induced climate change, the frequency and magnitude of such events are predicted to increase in a warming world.

In recently revised estimates scientists conclude that if greenhouse emissions (produced mainly by burning fossil fuels) are not curtailed, the Earth's average surface temperatures may increase from 2.7 degrees to nearly 11 degrees Fahrenheit by the end of the century, substantially more than the estimated 6.3 degrees in a 1995 report.*

The good news is, however, that the majority of experts believe that significant reductions in net greenhouse gas emissions are feasible due to an extensive array of policy and technological measures in the energy supply, energy demand, and agricultural and forestry sectors. In addition, the projected adverse effects of climate change on socioeconomic and ecological systems can, to some degree, be reduced through proactive adaptation measures.

Citizens of the United States of America have a particular obligation to address the threat of climate change. The US with 4.5 of the world population emits nearly 30 percent of the world's greenhouse gases.

WHEREAS the church, as the people of God, is called to be environmentally responsible in caring for God's creation as God's gift; and

WHEREAS the Annual Conference statement *Creation Called to Care* challenges us to take seriously our role as stewards of the Earth and to work for renewal of creation; and

WHEREAS our vastly increased use of fossil fuels has the potential to bring about irreversible changes in the climate and immense suffering for the poor and for people living in the coastal areas around the world;

THEREFORE BE IT RESOLVED that the Church of the Brethren General Board, meeting in New Windsor, Maryland on March 10-13, 2001, affirm the following principles:

1. Human societies must learn to draw on energy sources in ways that do not damage ecosystems or compromise the capacity of the Earth to meet the needs of current or future generations;

2. The generation and use of energy must be determined primarily by the needs of all people for a good quality of life, placing priority on appropriate and accessible energy for the world's poor;

3. Compliance with international trade agreements should not be given precedence over compliance with international environmental agreements or prevent the US from adopting measures to reorient its energy policy;

4. Energy policy in the US should be based on ethical principles of respect for and justice within the One Earth Community, focusing not on expanding supply through mega-projects but on managing the demand and development of renewable, alternative sources. Specifically, the US should:

   a. move beyond its dependence on high carbon fossil fuels that produce emissions leading to climate change;

   b. concentrate on reducing carbon dioxide emissions within the US and not rely on mechanisms such as emission trading with other countries to meet our targets for emission reductions under international agreements;

c. reduce our reliance on nuclear power, a technology for which there are still unresolved problems such as the safe disposal or safe storage of high level waste of nuclear reactors,

d. manage demand through high priority on conservation and energy efficiency,

e. significantly increase research and development into such renewable energy sources as solar, wind, biomass, etc.,

f. support development and utilization of appropriate technologies for small-scale, decentralized energy systems, and

g. provide necessary support for individuals, families, and communities adversely affected by a transition away from fossil fuels, nuclear power, and large-scale hydro in order to allow for alternative economic development, retraining, relocation, etc.

As members of the Church of the Brethren, we are encouraged to reduce our reliance on fossil fuels; to build and renovate our homes, church facilities, and camp structures to be energy efficient; to initiate new programs of energy conservation and awareness; to use public transportation, carpooling, and teleconferencing to reduce fossil fuel consumption; to become ecologically aware by using diets and products that consume less energy in production, transportation, packaging, and use; to separate and recycle household goods and to reduce waste and toxic materials.

> *God redeems us to live in community with the created Earth. We will care for God's Earth in ways that are sustainable* (Statement on *Simple Life,* 1996 Annual Conference minutes, p. #326).

Be it further resolved that the General Board ask staff to give priority to the issue of global warming/climate change; and provide models and educational resources for congregations, institutions, and members to study the issues; and take commensurate actions.

CHURCH OF THE BRETHREN POLICY BASE FOR THE RESOLUTION

Ecology (Annual Conference, 1971) *Energy Crisis* (General Board, 1973) *Concern on the Use of Energy and Resources* (General Board, 1975) *Justice and Nonviolence* (Annual Conference, 1977) *Christian Lifestyle* (Annual Conference, 1980) *A Quest for Order* (Annual Conference, 1987) *Global Warning and Atmospheric Degradation* (General Board, 1991) *Creation: Called to Care* (Annual Conference, 1991) *Simple Life* (Annual Conference, 1996)

* From the Third Assessment Report (January 2001) of the Intergovernmental Panel on Climate Change.

Global Climate Change

A Plea for Dialogue, Prudence, and the Common Good

(abridged)

Can we remain indifferent before the problems associated with climate change? Can we disregard the growing phenomenon of "environmental refugees," who are forced by the degradation of their natural habitat to forsake it.... in order to face the dangers and uncertainties of forced displacement?

— Pope Benedict XVI
2010 World Day of Peace

Introduction

As people of faith, we are convinced that "the earth is the Lord's and all it holds" (Ps 24:1). Our Creator has given us the gift of creation: the air we breathe, the water that sustains life, the fruits of the land that nourish us, and the entire web of life without which human life cannot flourish. All of this God created and found "very good." We believe our response to global climate change should be a sign of our respect for God's creation.

The continuing debate about how the United States is responding to questions and challenges surrounding global climate change is a test and an opportunity for our nation and the entire Catholic community. As bishops, we are not scientists or public policymakers. We enter this debate not to embrace a particular treaty, nor to urge particular technical solutions, but to call for a different kind of national discussion.

Much of the debate on global climate change seems polarized and partisan. Science is too often used as a weapon, not as a source of wisdom. Various interests use the airwaves and political process to minimize or exaggerate the challenges we face. The search for the common good and the voices of poor people and poor countries sometimes are neglected.

At its core, global climate change is not about economic theory or political platforms, nor about partisan advantage or interest group pressures. It is about the future of God's creation and the one human family. It is about protecting both "the human environment" and the natural environment. It is about our human stewardship of God's creation and our responsibility to those who come after us. With these reflections, we seek to offer a word of caution and a plea for genuine dialogue as the United States and other nations face decisions about how best to respond to the challenges of global climate change.

As Catholic bishops, we seek to offer a distinctively religious and moral perspective

Because of the blessings God has bestowed on our nation and the power it possesses, the United States bears a special responsibility in its stewardship of God's creation to shape responses that serve the entire human family. As pastors, teachers, and citizens, we bishops seek to contribute to our national dialogue by examining the ethical implications of climate change. We offer some themes from Catholic social teaching that could help to shape this dialogue, and we suggest some directions for the debate and public policy decisions that face us.

As Catholic bishops, we seek to offer a distinctively religious and moral perspective to what is necessarily a complicated scientific, economic, and political discussion. Ethical questions lie at the heart of the challenges facing us. As Catholic bishops, we accept the conclusions of the Intergovernmental Panel on Climate Change (IPCC).... Over the past few decades, the evidence of global climate change and the scientific consensus about the human impact on this process have led many governments to conclude that they need to invest time, money, and political will to address the problem.

Therefore, we especially want to focus on the needs of the poor, the weak, and the vulnerable in a debate often dominated by more powerful interests. Inaction and inadequate or misguided responses to climate change will likely place even greater burdens on already desperately poor peoples. Action to mitigate global climate change must be built upon a foundation of social and economic justice that does not put the poor at greater risk or place disproportionate and unfair burdens on developing nations.

* Scientific Knowledge and the Virtue of Prudence

The virtue of prudence is paramount in addressing climate change. This virtue is vital to the moral health of the larger community. It allows us to discern what constitutes the common good in a given situation. Prudence requires a deliberate and reflective process that aids in the shaping of community conscience. Prudence not only helps us identify the principles in a given issue, but also moves us to adopt courses of action to protect the common good. Prudence is not... simply a cautious and safe approach to decisions. Rather, it is a thoughtful, deliberate, and reasoned basis for taking action to achieve a moral good.

In facing climate change, what we already know requires a response; it cannot be easily dismissed. Significant levels of scientific consensus... justifies taking action intended to avert potential dangers. In other words, if enough evidence indicates that the present course of action could jeopardize humankind's well-being, prudence dictates taking mitigating or preventative action.

According to the IPCC, significant delays in addressing climate change compound the problem and make future remedies more difficult, painful, and costly. On the other hand, prudent actions today can potentially improve the situation over time, avoiding more sweeping action in the future.

* Climate Change and Catholic Social Teaching

We face two central moral questions: How are we to be stewards of creation in an age when we may alter that creation, perhaps irrevocably? How can we... exercise stewardship in a way that respects and protects the integrity of God's creation and provides for the common good? Catholic social teaching provides several themes to help answer these questions.

* The Universal Common Good

Global climate is by its very nature a part of the planetary commons. The earth's atmosphere encompasses all people, creatures, and habitats. The melting of ice sheets and glaciers, the destruction of rainforests, and the pollution of water in one place can have environmental impacts elsewhere. As Pope John Paul II has said, " *We cannot interfere in one area of the ecosystem without paying due attention to the consequences of such interference in other areas and to the well being of future generations.*" Responses to global climate change should reflect our interdependence and common responsibility for the future of our planet.

* Stewardship of God's Creation

Stewardship requires... careful protection of the environment... True stewardship requires changes in human actions. Our religious tradition has always urged restraint and

moderation in the use of material goods....
We must not allow our desire to possess more
material things overtake our concern for the
basic needs of people and the environment.
Changes in lifestyle based on moral virtues can
ease the way to a sustainable economy in which
sacrifice will no longer be an unpopular concept.
... A renewed sense of sacrifice and restraint
could make an essential contribution to
addressing global climate change.

* Protecting the Environment for Future Generations

The common good calls us to extend our
concern to future generations. Climate change
poses the question "What does our generation
owe to generations yet unborn?" As Pope John
Paul II has written, "there is an order in the
universe which must be respected. The human
person, endowed with the capability of choosing
freely, has a grave responsibility to preserve this
order for the well-being of future generations."

*According to the IPCC, delays
in addressing climate change
compound the problem and
make future remedies more
difficult, painful, and costly.*

Passing along the problem of global
climate change to future generations as a result
of our delay, indecision, or self-interest would
be easy. But we cannot leave this problem for
the children of tomorrow. As stewards of their
heritage, we have an obligation to respect their
dignity and to pass on their natural inheritance,
so that their lives are protected and, if possible,
made better than our own.

* Population and Authentic Development

Population and climate change should be
addressed from the perspective of a concern
for protecting human life, caring for the
environment, and respecting cultural norms
and the religious faith and moral values of
peoples.

* Caring for the Poor and Issues of Equity

Working for the common good requires us to
promote the flourishing of all human life and
all of God's creation.... No strategy to confront
climate change will succeed without the
leadership and participation of the United
States and other industrial nations. But any
successful strategy must include participation
from those most affected and least able to bear
the burdens. Developing and poorer nations
must have a place at the table.

* Public Policy Debate and Future Directions

The stronger and richer nations must have a
sense of moral responsibility for other nations,
so that an international system may be
established which will rest on the foundation of
the equality of all peoples and the necessary
respect for their legitimate differences.

CONCLUSION

As people of religious faith, we bishops believe
that the atmosphere that supports life on earth
is a God-given gift, one we must respect and
protect. If we harm the atmosphere, we
dishonor our Creator and the gift of creation.
The values of our faith call us to humility,
sacrifice, and a respect for life and the natural
gifts God has provided.

Pope John Paul II reminds us that
"respect for life and for the dignity of the
human person extends also to the rest of
creation, which is called to join man in praising
God." In that spirit, we Catholic bishops call
for prudent and constructive action to protect
God's precious gift of the earth's atmosphere
with a sense of genuine solidarity and justice
for all God's children.

– Msgr. William P. Fay
General Secretary
U.S. Catholic Conference of Bishops

General Board of
CHURCH & SOCIETY
of The United Methodist Church

Global Warming and Energy

The earth lies polluted under its inhabitants; for they have transgressed laws, violated the statutes, broken the ever-lasting covenant.

- Isaiah 24, NRSV

The crisis facing God's earth is clear. We, as stewards, have failed to live into our responsibility to care for creation and have instead abused it in ways that now threaten life around the planet.

The scientific consensus is clear that human activities are leading to a warming of the surface temperatures of the planet and the effects of this warming are being felt now and will be felt more intensely in years to come.

As a matter of stewardship and justice, Christians must take action now to reduce global warming pollution and stand in solidarity with our brothers and sisters around the world whose land, livelihood and lives are threatened by the global climate crisis.

"The scale of human activity has grown so large that it now threatens the planet itself. Global environmental problems have become so vast they are hard to comprehend. ... The vast majority of scientific evidence suggests that carbon dioxide from fossil fuels has already caused a measurable warming of the globe. Confronted with the massive crisis of the deterioration of God&'s creation and faced with the question of the ultimate survival of life, we ask God's forgiveness for our participation in this destruction." (2004 Book of Resolutions, "Environmental Justice for a Sustainable Future," ¶7)

"A transition to energy efficiency and renewable energy sources will combat global warming, protect human health, create new jobs, and ensure a secure, affordable energy future." (2004 Book of Resolutions, "Energy Policy Statement," ¶5)

"The U.S. must move beyond its dependence on high carbon fossil fuels that produce emissions leading to climate change and ratify the Kyoto Protocol under the U.N. Framework Convention on Climate Change." (2004 Book of Resolutions, "U.S. Energy Policy and United Methodist Responsibility," Scripture ref.: Isa 24:4-5 and 2 Chron. 7:14

The General Board of Church and Society (GBCS) advocates for the United States to adopt mandatory global warming emissions reductions and re-engage in the global dialogue and international framework for combating this clear and present danger. In collaboration with ecumenical and interfaith allies, GBCS supports a climate and energy campaign that focuses on both state and federal action. In addition, GBCS provides educational and worship resources to bring this issue into the life of the church.

United Methodist Church

FAITH PRINCIPLES ON GLOBAL WARMING

Justice: Strive for justice and acknowledge that global warming's societal impact already falls... most heavily on the people around the world who are least able to mitigate the impacts—poor and vulnerable populations in the U.S. and in developing countries. As a leading industrialized nation that has disproportionately contributed to greenhouse gas emissions, it is incumbent upon us to rectify this injustice.

To reach our goal of justice, we require that legislation:

o Include mechanisms that mitigate the impacts of global warming particularly for vulnerable populations in the U.S. and abroad.

o Prevent further harm to human health and all of God's creation by utilizing clean energy sources when addressing global warming and carbon pollution.

o Focus on a fair and equitable distribution of benefits and costs among people, communities, and nations, and rectify the disproportionate impact that low-income communities have and will experience as the climate continues to change.

o Enable our brothers and sisters in poverty to have economic independence and stability and to eliminate the devastating impacts [of] global warming....

o Take action now to avoid placing the burden of carbon reduction on our children's children.

o Endorse policies that place a high priority on allowing all people to live in God's abundance and with dignity by ensuring that basic human needs and justice are not adversely impacted by the effects of global warming....

Stewardship: Heed the call to be faithful stewards of God's creation by limiting the impacts of global warming on God's Earth. Already, global warming has damaged the balance of creation, including increasing... threatened species, long-term drought, and melting Arctic ice. To reach our stewardship goal, we require that legislation:

o Follow recognized scientific recommendations to protect all of God's creation and prevent catastrophic damage to God's Earth and people. Legislation must include comprehensive, mandatory, and aggressive emission reductions that limit the increase in temperature to 2 degrees Celsius or less. Legislation should reduce U.S. carbon emissions to reach a 15-20 percent reduction in carbon by 2020 with a vision to achieve carbon emissions that are 80 percent of 2000 levels by 2050.

o Avoid catastrophic global warming, which would devastate God's creation, put more pressure on disaster relief responses, and endanger the future of the planet.

o Call on major emitters to take responsibility for their actions and work to significantly reduce their carbon emissions.

Sustainability: Ensure that efforts to curb global warming prevent further environmental and societal tragedies. As people of faith we are guided by the value of sustainability. Sustainability requires that we enable biological and social systems that nurture and support life not be depleted or poisoned. To reach our goal of sustainability, we require that legislation:

o Maintain God's good creation by preventing policies that place the burden of our lifestyles on one aspect of creation and encouraging policies that sustain and restore vibrant eco-systems with economic justice so that communities of life can flourish for generations to come.
o Respond to global warming in a way that reflects the interdependence of all of God's creation.
o Support energy sources that are renewable, clean, and avoid destruction of God's creation.

Sufficiency: In a world of finite resources, for all to have enough requires that those among us who have more than enough will need to address our patterns of acquisition and consumption. We can not achieve significant reductions in global warming emissions unless we make changes in our lifestyles and particularly in our energy consumption. To support the goal of sufficiency, legislation must:

o Encourage energy conservation in our homes, our communities, and our places of worship.
o Encourage energy conservation in national transportation and distribution systems and commercial enterprises.
o Encourage the federal government to lead through research and example in the practice and implementation of energy conservation.

Social Principles of The United Methodist Church

2009-2012

D) Global Climate Stewardship—We acknowledge the global impact of humanity's disregard for God's creation. Rampant industrialization and the corresponding increase in the use of fossil fuels have led to a buildup of pollutants in the earth's atmosphere.

These "greenhouse gas" emissions threaten to alter dramatically the earth's climate for generations to come with severe environmental, economic, and social implications.

The adverse impacts of global climate change disproportionately affect individuals and nations least responsible for the emissions.

We therefore support efforts of all governments to require mandatory reductions in greenhouse gas emissions and call on individuals, congregations, businesses, industries, and communities to reduce their emissions.

The Evangelical Climate Initiative

CLIMATE CHANGE
An Evangelical Call to Action (abridged)

Preamble

AS AMERICAN EVANGELICAL CHRISTIAN LEADERS, we recognize both our opportunity and our responsibility to offer a biblically based, moral witness that can help shape public policy in the most powerful nation on earth, and therefore contribute to the well-being of the entire world. Whether we will enter the public square and offer our witness there is no longer an open question. We are in that square, and we will not withdraw.

We are proud of the evangelical community's long-standing commitment to the sanctity of human life. But we also offer moral witness on many issues.... While individuals and organizations can concentrate on certain issues, we are not a single-issue movement. We seek to be true to our calling as Christian leaders, and above all faithful to Jesus Christ our Lord.

Over the last several years many of us have engaged in study, reflection, and prayer related to climate change (often called "global warming"). For most of us, this has not been a priority. Indeed, many of us have required considerable convincing before becoming persuaded that climate change is a real problem and that it ought to matter to Christians. But now we have seen enough to offer the following moral argument related to human-induced climate change.

Claim 1: Human-Induced Climate Change is Real

Since 1995 there has been general agreement among those in the scientific community most seriously engaged with the issue that climate change is happening and is being caused by human activities, especially the burning of fossil fuels.

Because all religious/moral claims about climate change are relevant only if climate change is real and human-induced, everything hinges on the scientific data. As evangelicals we have hesitated to speak until we could be certain of the science of climate change, but the signatories now believe that the evidence demands action.

o The Intergovernmental Panel on Climate Change (IPCC), the world's most authoritative scientists on global warming, has been studying this since the 1980s. (From 1988-2002 the IPCC's assessment of the climate science was Chaired by Sir John Houghton, a devout evangelical Christian.) It has documented the steady rise in global temperatures over the last fifty years, projects that global temperature will continue to rise, and attributes "most of the warming" to human activities.

o The U.S. National Academy of Sciences, as well as all other scientific Academies (Great Britain, France, Germany, Japan, Canada, Italy, and Russia), has concurred with these judgments.

o In a 2004 report, and at the 2005 G-8 Summit, the Bush Administration acknowledged the reality of climate change and the likelihood that human activity is the cause. In the face of the breadth and depth of this scientific and governmental concern, only a small percentage of which is noted here, we are convinced that evangelicals must engage this issue without any further lingering over the basic reality of the problem or humanity's responsibility to address it.

Claim 2: The Consequences of Climate Change Will Be Significant

The earth's natural systems are resilient, but not infinitely so, and human civilizations are remarkably dependent on ecological stability, and well-being. It is easy to forget this until that stability and well-being are threatened. Even small rises in global temperatures will have such impacts as sea level rise; more frequent heat waves, droughts, and extreme weather events such as torrential rains and floods; increased tropical diseases in temperate regions; and more intense hurricanes. It could lead to significant reduction in agricultural output, especially in poor countries. Low-lying regions could find themselves under water.

Each of these impacts increases the likelihood of refugees from flooding or famine, violent conflicts, and international instability, which could lead to more security threats to our nation. Poor nations and poor individuals have fewer resources available to cope with major threats. Millions of people could die because of climate change.

The consequences of global warming will hit the poor the hardest, in part because those areas likely to be significantly affected first are in the poorest regions of the world. Millions of people could die in this century because of climate change, most of them our poorest global neighbors.

Claim 3: Christian Moral Convictions Demand Our Response

While we cannot here review the full range of relevant biblical convictions related to care of creation, we emphasize the following points

o Christians must care about climate change because we love God the Creator and Jesus our Lord, through whom and for whom the creation was made. This is God's world, and any damage that we do to God's world is an offense against God Himself (Gen. 1; Ps. 24; Col. 1).

o Christians must care about climate change because we are called to love our neighbors, and protect and care for the least of these as though Jesus Christ (Mt. 22:34-40; Mt. 7:12; Mt. 25:31-46).

o Christians, noting the fact that the climate change problem is human induced, are reminded that when God made humanity he commissioned us to exercise stewardship over the earth and its creatures. Climate change is the latest evidence of our failure to exercise proper stewardship, and constitutes a critical opportunity for us to do better (Gen. 1:26-28).

Love of God, love of neighbor, and the demands of stewardship are more than enough reason for evangelical Christians to respond to the climate change problem with moral passion and concrete action.

Claim 4: The Need To Act Is Urgent

Governments, businesses, churches and individuals have a role to play in addressing climate change – starting now. The basic task for all of the world's inhabitants is to find ways now to begin to reduce the carbon dioxide emissions from the burning of fossil fuels that are the primary cause of human-induced climate change.

There are several reasons for urgency. First, deadly impacts are being experienced now. Second, the oceans only warm slowly, creating a lag in experiencing the consequences. Much of the climate change to which we are already committed will not be realized for several decades. The consequences of the pollution we create now will be visited upon our children and grandchildren. Third, we are making long-term decisions today that will determine how much carbon dioxide we will emit in the future, such as whether to purchase energy efficient vehicles and appliances that will

last for 10-20 years, or whether to build more coal-burning power plants that last for 50 years rather than investing more in energy efficiency and renewable energy.

Numerous positive actions to prevent climate change are being implemented by state and local governments, churches, businesses, and individuals. These commendable efforts focus on energy efficiency, renewable energy, low CO_2-emitting technologies, and the use of hybrid vehicles. These efforts can save money, save energy, reduce global warming pollution as well as air pollution that harm human health, and eventually pay for themselves.

Finally, while we must reduce our global warming pollution to help mitigate the impacts of climate change, as a society and as individuals, we must also help the poor adapt to the significant harm that global warming will cause.

Conclusion

WE THE UNDERSIGNED PLEDGE TO ACT ON THIS DOCUMENT. We will not only teach the truths communicated here but seek ways to implement the actions that follow from them. In the name of Jesus Christ our Lord, we urge all who read this declaration to join us in this effort.

(Institutional affiliation is listed for identification purposes only)

Rev. Dr. Leith Anderson, National Association of Evangelicals, Eden Prairie, MN

Robert Andringa, Ph.D., Council for Christian Colleges and Universities, Vienna, VA

Rev. Jim Ball, Ph.D., Evangelical Environmental Network; Wynnewood, PA

Commissioner Todd Bassett, Natl. Commander, The Salvation Army; Alexandria, VA

Dr. Jay A. Barber, Jr., President, Warner Pacific College, Portland, OR

Gary P. Bergel, President, Intercessors for America; Purcellville, VA

David Black, Ph.D., President, Eastern University, St. Davids, PA

Bishop Charles Blake, Sr., West Angeles Church of God in Christ, Los Angeles, CA

Rev. Dan Boone, President, Trevecca Nazarene University, Nashville, TN

Bishop Wellington Boone, The Father's House & Wellington Ministries, Norcross, GA

Rev. Dr. Peter Borgdorff, Christian Reformed Church, Grand Rapids, MI

H. David Brandt, Ph.D., President, George Fox University, Newberg, OR

Rev. George K. Brushaber, Ph.D., President, Bethel University; St. Paul, MN

Rev. Dwight Burchett, President, No. Calif. Assoc. of Evangelicals; Sacramento, CA

Gaylen Byker, Ph.D., President, Calvin College, Grand Rapids, MI

Rev. Dr. Jerry B. Cain, President, Judson College, Elgin, IL

Rev. Dr. Clive Calver, Former President, World Relief; Bethel, CT

R. Judson Carlberg, Ph.D., President, Gordon College, Wenham, MA

Rev. Dr. Paul Cedar, Chair, Mission America Coalition; Palm Desert, CA

David Clark, Ph.D., President, Palm Beach Atlantic University; CEO, Nat. Rel. Broadcasters; West Palm Beach, FL

Rev. Paul de Vries, Ph.D., President, New York Divinity School; New York, NY

Larry R. Donnithorne, President, Colorado Christian University, Lakewood, CO

Blair Dowden, Ed.D., President, Huntington University, Huntington, IN

Rev. Robt Dugan, Jr., VP of Gov. Affairs, Natl. Ass. of Evangelicals, Palm Desert, CA

Craig Hilton Dyer, President, Bright Hope International, Hoffman Estates, IL

D. Merrill Ewert, Ed.D., President, Fresno Pacific University, Fresno, CA

Rev. Dr. LeBron Fairbanks, Mount Vernon Nazarene University, Mount Vernon, OH

Rev. Myles Fish, President/CEO, International Aid, Spring Lake, MI

Rev. Timothy George, Ph.D., Dean, Beeson Divinity School, Birmingham, AL

Rev. Michael J. Glodo, Stated Clerk, Evangelical Presbyterian Church , Livonia , MI

Rev. Dr. Jeffrey E. Greenway, Asbury Theological Seminary, Wilmore, KY

among hundreds of other signatories

Presbyterian Church (USA)

Policy Brief

Climate Change and Energy

(Excerpted and expanded from the Christian and Citizen Election Year Resource)

Climate Change

General Assemblies have repeatedly expressed concern about the impact of climate change on God's creation. The 2008 statement, "The Power to Change: U.S. Energy Policy and Global Warming," included many policy recommendations that intimately link the nation's energy policy with its climate change legislation.

The Assembly began with international policy recommendations for the U.S., ...asserting that the U.S., "which has historically produced more greenhouse gasses than any other country, and which is currently responsible for over a fifth of the world's annual emissions," must accept its "moral responsibility" to be a world leader in resolving the crisis. The Assembly, therefore, reaffirmed the call of four previous Assemblies for the U.S. to sign the Kyoto Protocol and urged the U.S. government to "repent of its efforts to block consensus and to work with the international community as it develops a binding agreement to replace the Kyoto Protocol...." Further, commissioners "reject[ed] the claim that all nations should shoulder an equal measure of the burden associated with mitigating climate change. Industrialized nations like the United States that have produced most of the emissions over the last three centuries deserve to shoulder the majority of the burden."

On recommendations for policy to regulate U.S. emissions, the "Presbyterian Church (U.S.A.) supports comprehensive, mandatory, and aggressive emission reductions that aim to limit the increase in Earth's temperature to 2 degrees Celsius or less from pre-industrial levels. Legislation should focus on the short-term goal of reducing U.S. greenhouse gas emissions 20 percent from 1990 levels by 2020, and 80 percent from 1990 levels by 2050." In order to do so, the Assembly called for the ,social and environmental costs related to greenhouse gas emissions "to be "internalize[d]" into the price of fossil fuel....
(http://www.pc-biz.org/Explorer.aspx?id=1537&promoID=10)

Climate Change and Poverty

The commissioners at the 218th General Assembly (2008) introduced the policy recommendations on climate change with a commitment to "stand with "the least of these" and advocate for the poor and oppressed in present and future generations who are often the victims of environmental injustice and who are least able to mitigate the impact of global warming that will fall disproportionately on them." When discussing government revenues that would be generated by a cap-and-trade system and/or carbon tax, the commissioners recommended that they be "utilized nationally to redress the regressive impact of higher energy prices on people who are poor... Internationally, the United States needs to contribute funds to help poorer nations adapt to the social dislocation and ecological devastation caused by global climate change."

Carbon Neutral Lives

In 2006, the 217th General Assembly passed a resolution, "strongly urging all Presbyterians to immediately make a bold witness by aspiring to live carbon neutral lives. (Carbon neutrality requires our energy consumption that releases carbon dioxide into the atmosphere be reduced and carbon offsets purchased to compensate for those carbon emissions that could not be eliminated.)" (*Minutes*, 2006, pp. 895-898)

U.S. Energy Policy

Commissioners to the 218th General Assembly (2008) approved a number of specific energy-related policy recommendations, including:

- "Shift subsidies and financial incentives toward industries specializing in renewable energy and energy efficiency and away from fossil fuel and nuclear power industries.

- "Adopt significantly increased efficiency standards for all energy consuming appliances, buildings, and vehicles, including CAFE standards for cars and trucks.

- "Mandate that an increasing percentage of the nation's energy supply be produced renewably and sustainably" and calling for a "20 percent national Renewable Energy Standard (RES) by 2020."

- "Remove market barriers for producers of renewable energy... [and] encourage decentralized and distributed power generation."

- "Place a moratorium on all new coal-fired and nuclear power plants until environmental concerns are addressed... It would be irresponsible to build new coal-fired power plants.... until it can be demonstrated that the carbon can be captured economically and sequestered permanently.

Limit New Fossil Fuel Exploration

The 218th General Assembly (2008) called for legislative and policy proposals that: "limit exploration and exploitation of new fossil fuel supplies to parts of the nation where this can be done without adverse damage to people and the environment. As the climate in the Arctic warms, it is doubtful that the economic benefits of drilling in the Arctic National Wildlife Refuge can outweigh the environmental damage that this will do to one of the nation's most beautiful and wild places. Another example of such a limit would be the ecological devastation associated with mountaintop removal mining in Appalachia."

(http://www.pc-biz.org/Explorer.aspx?id=1537&promoID=10)

Energy as National Security

The 218th General Assembly (2008) called on the U.S. to revise national security policies. "Decrease attempts to control oil resources owned by other nations and the profligate use of energy supplies to enforce inevitably temporary as well as massively tragic military interventions. Increase the authority of science-based international standards for addressing the issue of global climate change. Strive to decouple nuclear power from nuclear weapons production so as not to encourage a new round of nuclear proliferation. "

(http://www.pc-biz.org/Explorer.aspx?id=1537&promoID=10)

National Council of Churches
That the world might believe

The Eco-Justice Working group

God's Earth is Sacred
An Open Letter to Church and Society

February 14, 2005   *   Abridged text

GOD'S CREATION DELIVERS UNSETTLING NEWS. Earth's climate is warming to dangerous levels; 90% of the world's fisheries have been depleted; coastal development and pollution are causing a decline in ocean health; over 95% of U.S. forests have been lost; and almost half of the population in the United States lives in areas that do not meet national air quality standards. In recent years, the profound danger has grown, requiring us as theologians and religious leaders to speak out and act with new urgency.

We are obliged to relate to Earth as God's creation "in ways that sustain life [and] provide for the needs of humankind. The religious community in the U.S. has addressed issues of ecology and justice.

To continue to walk the current path of ecological destruction is not only folly; it is sin. As voiced by Ecumenical Patriarch Bartholomew, who has taken the lead among religious leaders in his concern for creation: "To commit a crime against the natural world is a sin. For humans to cause species to become extinct and to destroy the biological diversity of God's creation . . . for humans to degrade the integrity of Earth by causing changes in its climate, by stripping the Earth of its natural forests, or for humans to contaminate the Earth's waters, its land, its air, and its life, with poisonous substances . . . these are sins." We have become un-Creators. Earth is in jeopardy at our hands.

The first step is to repent of our sins.... This repentance of our social and ecological sins will acknowledge the responsibility that falls to [us]. Though only 5% of the planet's human population, we produce one-quarter of the world's carbon emissions, consume a quarter of its natural riches, and perpetuate inequities at home and abroad. We do not own the planet and we cannot transcend its requirements for regeneration.... We have not listened to the Maker of Heaven and Earth.

The second step is to pursue a new journey.... We can share in that renewal by clinging to God's promise to restore and fulfill all that God creates and by walking a new path.... We firmly believe that addressing the degradation of God's sacred Earth is the moral assignment of our time.

Ecological Affirmations of Faith

We stand with awe and gratitude as members of God's good creation. We rejoice in the splendor and mystery of countless species, and the interdependence of all that God makes. We believe that the Earth is home for all and that it has been created intrinsically good (Genesis 1).

We lament that humans are shattering the gifts of this web of life, ignoring our responsibility for the well being of all life, while destroying species at a rate never before known in human history.

We believe that the Holy Spirit animates all creation, breathes in us, and can empower us to participate in working toward the flourishing of Earth's community of life. We believe that the people of God are called to forge ways that enable socially just and ecologically sustainable communities to flourish for generations to come. And we believe in God's promise to fulfill all of creation, anticipating the reconciliation of all (Col. 1:15), in accordance with God's promise (II Pet 3:13).

We lament that we have rejected this vocation, have distorted our God-given abilities to destroy ecosystems and human communities rather than to protect, strengthen, and nourish them.

We confess that instead of proclaiming salvation through our lives and worship, we have abused and exploited the Earth and people on the margins of power and privilege, altering climates, extinguishing species, and jeopardizing Earth's capacity to sustain life as we know and love it.

We believe that the created world is sacred - a revelation of God's power and gracious presence filling all things. This sacred quality of creation demands moderation and sharing.... We cling to God's promise to restore, renew, and fulfill all that God creates.

Guiding Norms for Church and Society

These affirmations imply a challenge that is also a calling: to fulfill our vocation as images of God, charged to "serve and preserve" the Garden (Gen 2:15). We affirm the following responsibility:

Justice-creating right relationships, both social and ecological, to ensure for all members of the Earth community the conditions required for their flourishing.

Sustainability - living within the bounds of planetary capacities, in fairness to both present and future generations of life. forces us to be responsible for the truly long-term impacts of our lifestyles.

Bioresponsibility - extending the covenant of justice to include all other life forms as beloved creatures of God and as expressions of God's presence, wisdom, power, and glory.

Humility - recognizing, as an antidote to arrogance, the limits of human knowledge, technological ingenuity, and moral character. We are not the masters of creation.

Generosity - sharing Earth's riches to promote and defend the common good in recognition of God's purposes for the whole creation and Christ's gift of abundant life.

Frugality - restraining economic production and consumption for the sake of eco-justice. Frugality connotes moderation, sufficiency, and temperance. It demands the careful conservation of Earth's riches.

Solidarity - acknowledging that we are bound together as a global community and bear responsibility for one another's well being. Our problems must be addressed with cooperative action at all levels...

A Call to Action: Healing the Earth and Providing a Just and Sustainable Society

We believe that caring for creation must undergird ... all dimensions of church ministries. Therefore, we call on our brothers and sisters in Christ, and all people of good will, to join us in:

We believe that one of the surest ways to gain this understanding is by listening to the most vulnerable: those who most suffer the consequences of our overconsumption, toxication, and hubris.

Integrating this understanding into our core beliefs and practices surrounding what it means to be "church," to be "human," to be "children of God." ... With this integrated witness we look to a revitalization of our human vocation and our churches' lives that revitalizes God's thriving Earth.

Advocating boldly with all our leaders on behalf of creation and its most vulnerable members. We must shed our complacency, denial, and fears and speak God's truth to power....

In Christ's name and for Christ's glory, we call out with broken yet hopeful hearts: join us in restoring God's Earth - the greatest healing work and moral assignment of our time. †

The Episcopal Church

A Pastoral Teaching on the Environment

The House of Bishops
Office of Public Affairs
September 20, 2011
(abridged)

WE, YOUR BISHOPS, BELIEVE THESE WORDS OF JEREMIAH describe these times and call us to repentance as we face the unfolding environmental crisis of the earth:

> *How long will the land mourn, and the grass of every field wither? For the wickedness of those who live in it the animals and the birds are swept away, and because people said, "He is blind to our ways.* (Jeremiah 12:4)

The mounting urgency of our environmental crisis challenges us... to confess "our self-indulgent appetites and ways," "our waste and pollution of God's creation," and "our lack of concern for those who come after us" (Ash Wednesday Liturgy, Book of Common Prayer, p. 268). It also challenges us to amend our lives and to work for environmental justice and for more environmentally sustainable practices.

Christians cannot be indifferent to global warming, pollution, natural resource depletion, species extinctions, and habitat destruction, all of which threaten life on our planet. Because so many of these threats are driven by greed, we must also actively seek to create more compassionate and sustainable economies that support the well-being of all God's creation.

One of the most dangerous and daunting challenges we face is global climate change.

We are especially called to pay heed to the suffering of the earth. The Anglican Communion Environmental Network calls to mind the dire consequences our environment faces: "We know that . . . we are now demanding more than [the earth] is able to provide. Science confirms what we already know: our human footprint is changing the face of the earth and because we come from the earth, it is changing us too. We are engaged in the process of destroying our very being....

This is the appointed time for all God's children to work for the common goal of renewing the earth as a hospitable abode for the flourishing of all life. We are called to speak and act on behalf of God's good creation.

Looking back to the creation accounts in Genesis, we see God's creation was *"very good,"* providing all that humans would need for abundant, peaceful life. In creating the world God's loving concern extended to the whole of it, not just to humans. And the scope of God's redemptive love in Christ is equally broad: the Word became incarnate in Christ not just for our sake, but for the salvation of the whole world. In the Book of Revelation we read that God will restore the goodness and completeness of creation in the "new Jerusalem." Within this new city, God renews and redeems the natural world rather than obliterating it. We now live in that time between God's creation of this good world and its final redemption: *"The whole creation has been groaning in labor pains until now; and not only the creation, but we ourselves, who have the first fruits of the Spirit, groan inwardly while we wait for . . . the redemption of our bodies"* (Romans 8:22-3).

Affirming the biblical witness to God's abiding and all-encompassing love for creation, we recognize that we cannot separate ourselves from the rest of the created order. The creation story itself presents the interdependence of all God's creatures in their wonderful diversity and fragility, and in their need of protection from dangers of many kinds. This is why the Church prays regularly for the peace of the whole world, for seasonable weather and an abundance of the fruits of the earth, for a just sharing of resources.... This includes our partner creatures: animals, birds, and fish who are being killed or made sick by the long-term effects of deforestation, oil spills, and a host of other ways in which we intentionally and unintentionally destroy or poison their habitat.

One of the most dangerous and daunting challenges we face is global climate change. This is, at least in part, a direct result of our burning of fossil fuels. Such human activities could raise worldwide average temperatures by three to eleven degrees Fahrenheit in this century. Rising average temperatures are already wreaking environmental havoc, and, if unchecked, portend devastating consequences for every aspect of life on earth.

The Church has always had as one of its priorities a concern for the poor and the suffering. Therefore, we need not agree on the fundamental causes of human devastation of the environment, or on what standard of living will allow sustainable development, or on the roots of poverty in any particular culture, in order to work to minimize the impact of climate change. It is the poor and the disadvantaged who suffer most from callous environmental irresponsibility. Poverty is both a local and a global reality. A healthy economy depends absolutely on a healthy environment.

> *Privileged Christians... need to move from a culture of consumerism to a culture of conservation and sharing.*

The wealthier nations whose industries have exploited the environment, and who are now calling for developing nations to reduce their impact on the environment, seem to have forgotten that those who consume most of the world's resources also have contributed the most pollution to the world's rivers and oceans, have stripped the world's forests of healing trees, have destroyed both numerous species and their habitats, and have added the most poison to the earth's atmosphere. We cannot avoid the conclusion that our irresponsible industrial production and consumption-driven economy lie at the heart of the current environmental crisis.

Privileged Christians in our present global context need to move from a culture of consumerism to a culture of conservation and sharing. The challenge is to examine one's own participation in ecologically destructive habits. Our churches must become places where we have honest debates about, and are encouraged to live into, more sustainable ways of living. God calls us to die to old ways of thinking and living and be raised to new life with renewed hearts and minds.

Although many issues divide us as people of faith, unprecedented ecumenical and interfaith cooperation is engaging the concern to protect our planet. And yet, efforts to stop environmental degradation must not be simply imposed from above. Those most affected must have a hand in shaping decisions....

Our current environmental challenges call us to ... repentance: we must turn ourselves around, and come to think, feel, and act in new ways. Ancient wisdom and spiritual disciplines from our faith offer deep resources to help address this environmental crisis. Time-honored practices of fasting, Sabbath-keeping, and Christ-centered mindfulness bear particular promise for our time.

Fasting disciplines and heals our wayward desires and appetites, calling us to balance our individual needs with God's will for the whole world. In fasting we recognize that human hungers require more than filling the belly. In God alone are our desires finally fulfilled. Commended in the Book of Common Prayer, fasting is grounded in the practices of Israel, taught by Jesus, and sustained in Christian tradition. The ecological crisis extends and deepens the significance of such fasting as a form of self-denial: those who consume more than their fair share must learn to exercise self-restraint so that the whole community of creation might be sustained. ...

The practice of Christ-centered mindfulness, that is, the habitual recollection of Christ, calls believers to a deepened awareness of the presence of God in their own lives, in other people, and in every aspect of the world around us. Such spiritual perception should make faithful people alert to the harmful effects of our lifestyles, attentive to our carbon footprint and to the dangers of overconsumption. It should make us profoundly aware of the gift of life and less prone to be ecologically irresponsible in our consumption and acquisition. In assuming with new vigor our teaching office, we, your bishops, commit ourselves to a renewal of these spiritual practices in our own lives, and invite you to join us in this commitment for the good of our souls and the life of the world. Moreover, in order to honor the goodness and sacredness of God's creation, we, as brothers and sisters in Christ, commit ourselves and urge every Episcopalian:

- To acknowledge the urgency of the planetary crisis in which we find ourselves, and to repent of any and all acts of greed, overconsumption, and waste that have contributed to it;

- To lift up prayers in personal and public worship for environmental justice, for sustainable development, and for help in restoring right relations both among humankind and between humankind and the rest of creation;

- To take steps in our individual lives, and in community, public policy, business, and other forms of corporate decision-making, to practice environmental stewardship and justice, including (1) a commitment to energy conservation and the use of clean, renewable sources of energy; and (2) efforts to reduce, reuse, and recycle, and whenever possible to buy products made from recycled materials;

- To seek to understand and uproot the political, social, and economic causes of environmental destruction and abuse;

- To advocate for a "fair, ambitious, and binding" climate treaty, and to work toward climate justice through reducing our own carbon footprint and advocating for those most negatively affected by climate change.

May God give us the grace to heed the warnings of Jeremiah and to accept the gracious invitation of the incarnate Word to live, in, with, and through him, a life of grace for the whole world, that thereby all the earth may be restored and humanity filled with hope. Rejoicing in your works, O Lord, send us forth with your Spirit to renew the face of the earth, that the world may once again be filled with your good things: the trees watered abundantly, springs rushing between the hills in verdant valleys, all the earth made fruitful, your manifold creatures, birds, beasts, and humans, all quenching their thirst and receiving their nourishment from you once again in due season (Psalm 104).

Evangelical Lutheran Church in America

God's Work, Our Hands

Statement on Climate Change

December 31, 2009

The earth dries up and withers, the world languishes and withers; the heavens languish together with the earth. The earth lies polluted under its inhabitants; for they have transgressed laws, violated the statutes, broken the everlasting covenant.   Isaiah 24:4-5

Climate change is one of the most pressing issues facing all of God's creation. The evidence is mounting that earth's climate is changing dramatically—sea levels are rising, rainfall patterns are changing, polar ice and glaciers are melting, weather disasters are increasing.

The impacts of these changes are already falling most heavily on people and nations struggling in poverty, those least able to adapt to changing conditions because they have few resources to do so. We see their stories in the news: drought in Africa; hurricanes in Central America; coastal flooding on small Pacific islands. People are hungry, forced to leave their homes and struggling to rebuild. Perhaps we are moved to act: to give money to a disaster appeal; to write a letter to our member of Congress urging them to provide international aid. But are we called to do more in the face of mounting evidence that this problem is growing worse?

The great paradox of climate change is that those least responsible for the emission of pollutants harmful to the earth will be most severely affected and least able to adapt to changing conditions.

Christ taught us to seek justice, to care for our neighbor and to provide special care and consideration for "the least of these" — those living in poverty.

Our response to climate change must heed this call to justice, particularly for those living in poverty around the world, who are least responsible for climate change and most likely to suffer greatly from its impacts if we do not act now.

As Christians, we are called to protect what God has created, and we are obligated to speak and to act in response to this growing crisis. As Lutherans, we are freed by the cross and resurrection of Jesus Christ to serve our neighbor and all of creation; and we are people of hope, who see the crisis facing our planet and are taking action.

What is climate change?

Climate change is real. A report released in 2007 by the Intergovernmental Panel on Climate Change (IPCC), a group of leading scientists from around the globe who have been studying data on climate for decades, confirmed that global average temperatures are growing warmer due to increasing levels of carbon dioxide and other heat-trapping gases in Earth's atmosphere, and that human use of fossil fuels is the main source of the increase in these gases. Every time we burn gasoline by driving a car, use electricity from coal- or gas-fired power plants, or heat our homes with oil or natural gas, we release carbon dioxide and "greenhouse" gases into the air. At normal levels greenhouse gases make Earth's temperatures moderate enough to support life, but at these increased, and human-caused levels, more and more of the sun's heat is trapped by our atmosphere and less escapes back into space. The increase in trapped heat changes the climate, causing altered weather patterns that can bring unusually intense precipitation, droughts and more severe storms. The United States produces 25 percent of the world's carbon dioxide emissions. Although China recently surpassed the United States as the largest emitter of carbon dioxide, our individual emissions remain the highest of any nation and we are the largest historic emitter of greenhouse gas.

What are the impacts of climate change?

In 2007, the IPCC painted a bleak picture of the future for God's creation and those struggling with hunger, poverty and disease. They predicted that a changing climate will increase food insecurity in places like Africa, where food is already scarce, while reversing progress made in the fight against hunger in regions like Latin America. Rising temperatures will increase freshwater scarcity in some regions and the spread of diseases such as malaria, dengue fever, and West Nile virus.... More severe storms and drought will lead to increased migration, and rising sea levels will likely lead to the permanent displacement of coastal communities and even entire small island nations. Increasing numbers of environmental migrants, coupled with greater competition for scarce resources among people and nations, are potentially destabilizing forces that cannot be ignored.

Some of these impacts can already be seen and measured. For example, smallholder farmers in western Nicaragua are struggling to adapt to increasingly unpredictable rainfall: where they once were able to plant and harvest two crops during the annual rainy season, they are now facing crop failure and hunger. Often there is too little rain, and crops die, or too much, and floods carry away seedlings. In several small communities, projects supported by ELCA World Hunger and The Lutheran World Federation are helping farmers develop irrigation systems to farm during the dry season. However, groundwater levels are dropping as well due to the unpredictable rainfall, and there is little money to invest in irrigation projects.

Families in these communities struggle to survive, and many men leave their communities — or even the country — to seek work as crops fail season after season. For these farmers climate change is a reality. As this example illustrates, climate change impacts fall most heavily on those living in poverty and populations who are dependent on their natural environment for their survival. Wealthy countries like the United States will better adapt to these changes, but as Hurricane Katrina proved in 2005, even in our country, low income people are highly vulnerable.

As Lutherans around the world accompany our neighbors in their journeys out of poverty, climate change presents new challenges and requires new strategies. For example, Lutheran World Relief's projects aim to improve peoples' lives and resiliency to climate change by increasing access to water, food security, reducing disaster risks and preventing and treating malaria.

How are we called to respond?

As a nation, we bear moral responsibility for this crisis. Scientists tell us that it is consumption of energy and resources in industrialized countries that is the primary cause of climate change.

Although the United States has only about five percent of global population, we consume 30 percent of the world's resources and create 30 percent of the world's waste. The U.S. is the largest historic emitter of carbon dioxide and we emit more carbon dioxide per person than any other nation. The planet is in danger unless we are willing to face this reality and to take responsibility for it by reducing our use of energy and our emissions of carbon dioxide and other greenhouse gases. If we don't, all of creation will suffer.

As a leading industrialized nation that has disproportionately contributed to greenhouse gas emissions, it is incumbent upon us to rectify this injustice through national legislation that meets the following goals:

Follow the recommendations of the scientific community to reduce greenhouse gas emissions.

Currently this means legislation must ensure that we do not increase the Earth's temperature by more than two degrees Celsius. Scientists say we can achieve this goal by reducing emissions by between 20 and 40 percent by 2020 and by 80 percent by 2050.

Protect those living in poverty in the U.S. from the impacts of climate change and climate legislation.

Legislation must ensure that low income Americans do not bear the disproportionate burden of increases in energy costs, must ensure that any increased costs do not push more people into poverty and must provide for those whose jobs are impacted by climate legislation.

Provide adaptation assistance for those living in poverty abroad.

Those living in the most vulnerable developing nations around the world bear little responsibility for greenhouse gas emissions and are already feeling the burden of climate change, with little ability to adapt. Through adaptation assistance, the U.S. can prevent the destruction of vulnerable communities around the world and help with climate relief.

As a nation we should work for a strong international agreement that requires all nations to reduce their emissions of carbon dioxide and other greenhouse gases. Any agreement should provide aid to developing nations to reduce their emissions even as they lift their people out of poverty, and help them adapt to the impacts of a changing climate on food, water and other basic human needs. And finally, as individuals we must examine our own role and responsibility for climate change. How can we act, in our homes, congregations and communities, to consume and waste less?

This resource is produced by Church in Society, a unit of the Evangelical Lutheran Church in America, dedicated to promoting peace, justice, and the care of creation in Christ's name in the global community.

Learn More: The **ELCA social statement** "Caring for Creation" states our concerns about climate change and its potential impacts on God's creation and calls us to act (*www.elca.org/environment).*

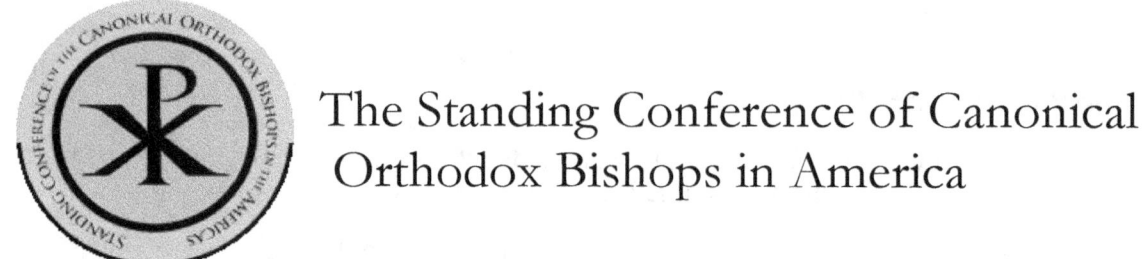

The Standing Conference of Canonical
Orthodox Bishops in America

GLOBAL CLIMATE CHANGE
A Moral and Spiritual Challenge

Resolution by The Standing Conference of Orthodox Bishops in America

May 25, 2007

THE CHURCH HAS ALWAYS KNOWN that human beings are dependent upon the grace of God through the world to nurture and sustain society. While God is the Source of all that we have, we humans share a responsibility to care for His creation and offer it back to Him in thanksgiving.... *"Thine own of thine own, we offer unto thee, in behalf of all and for all."* The action of returning creation back to God in gratitude summarizes the commands that God gave us in Genesis.

In our day, society has failed to remember these holy mandates.... As a people, we have forgotten God and our mandated responsibilities.... Instead of receiving the gifts of God as He would bestow them, we heedlessly take from the earth and waste its resources.... The pollution and degradation of the world is directly related to the pollution and the degradation of our hearts.

Faithful to the responsibility that we have been given, it is prudent to listen to the world's scientific leaders as they describe changes in the world's climate, changes that are already being experienced throughout the world. In Alaska, the average temperature has risen by 7°, causing glaciers to retreat.... In Florida, Hawaii and the islands of the Caribbean, coral reefs are dying. In ocean waters... higher temperatures result in lower concentrations of plankton, reducing a food source for fish and birds, and ultimately, for humans. Across the West, a modest increase in temperature has contributed to a six-fold increase in forest fires.... Previously distant diseases, such as West Nile virus and dengue fever, are appearing as a direct result of rising temperatures.

These are signs of a changing climate. While the world's climate has also under-gone changes in past centuries, three considerations make the current changes unprecedented:

* The rapid extent of temperature increase is historically unparalleled.

* The human role in changing the climate is unique today.

* The impact that climate change will exert upon society is great and diverse, including conditions which disrupt the lives and livelihoods of people on an unprecedented scale.

These changes are the result of increases of "greenhouse gases" in the atmosphere. These are produced by the burning of gasoline, coal and other fossil fuels. Among the consequences, the atmosphere and the oceans are warming; wind and rainfall patterns are changing; sea levels are rising. Forces of climate change also increase the acidity of the oceans; raise the ferocity of storms...; cause droughts and heat waves to become more intense; and, they disrupt normal agriculture. The conditions we observe now are only the early alterations to our climate. Larger and more disruptive changes will result unless we reduce the forces causing climate change.

It should be clear that immediate measures must be taken. If we fail to act now, the changes already underway will intensify and create catastrophic conditions. A contributing cause of these changes is a lifestyle that contains unintended... destructive side effects. It may be that no person intends to harm the environment, but excessive use of fossil fuels is destroying creation's life.

As Church leaders, it is our responsibility to speak to this condition as it represents a grave moral and spiritual problem.

> *Immediate measures must be taken. If we fail to act now, the changes already underway will intensify and create catastrophic conditions.*

Therefore, we... emphasize the seriousness and the urgency of the situation. To persist in a path of excess and waste, at the expense of our neighbors and beyond the capability of the planet to support the lifestyle responsible for these changes, is not only folly; it jeopardizes the survival of God's creation.... Not only is [this] sinful; it is no less than suicidal.

But there is hope. Society can alter its behavior and avoid the more serious consequences of climate change. In order to make the required changes, we are called to pray for a change in our personal attitudes and habits, in spite of any inconvenience. The issue is not merely our response to climate change, but our failure to obey God. We must live in a manner that is consistent with what we believe. At minimum, this means caring about the effect of our lives upon our neighbors, respecting the natural environment, and demonstrating a willingness to live within the means of our planet. Such a change will invariably require reduction in our consumption of fossil fuels as well as acceptance of alternative energy sources such as solar or wind power, and other methods that minimize our impact upon the world. We can do these things, but it will require intentional effort.

We must learn all that we can about climate change. We must set an example in the way that we live, informing others about this threat. We must discuss this with fellow parishioners and we must raise the issue before public officials. Each of us can   do something.

At every Divine Liturgy, we pray for seasonable weather. Let us enter into this prayer and amend our lives... If we can do this, God willing, we may live and flourish. This is not optional.... The Scriptures tell us if we destroy the earth, God will destroy us (see Rev. 11:18).

Let us all recall the commands of God regarding our use of the earth. And let us responsibly discern the right, holy and proper way to live in this time of change and challenge.

The Southern Baptist Convention

Declaration on
the Environment and Climate Change

A Minority Assessment

March 10, 2008
(abridged)

Preamble

God's great blessings on our denomination bestow upon us a great responsibility to offer a biblically-based, moral witness that can help shape individual behavior and public policy. Southern Baptists have always championed faith's challenges, and we perpetuate our heritage through this initiative. We are proud of our deep commitments to moral issues like the sanctity of human life and biblical definitions of marriage. We will never compromise our convictions nor attenuate our advocacy on these matters.... However, we are not a single-issue body. We offer moral witness on many issues. We seek to be true to our calling as Christian leaders, but above all, faithful to Jesus Christ our Lord. We have recently engaged in study, reflection and prayer related to the challenges presented by climate change. These things have not always been treated with pressing concern as major issues. Indeed, some of us have required considerable convincing before becoming persuaded that these are real problems that deserve our attention. But now we have seen and heard enough to be persuaded that these issues are among the current era's challenges that require a unified moral voice.

We believe our current denominational engagement with these issues have often been too timid, failing to produce a unified moral voice. Our cautious response to these issues in the face of mounting evidence may be seen as uncaring, reckless and ill-informed. We can do better. To abandon these issues is to shirk from our responsibility to be salt and light. The time for timidity regarding God's creation is no more.

Therefore, we offer these four statements for consideration, beginning with our fellow Southern Baptists, and urge all to follow by taking appropriate actions. May we find ourselves united as we contend for the faith that was delivered to the saints once for all. *Laus Deo!*

1. Humans Must Care for Creation and Take Responsibility for Environmental Degradation

There is undeniable evidence that the earth can be damaged by human activity and that people suffer as a result. When this happens, it is especially egregious because creation serves as a revelation of God's presence, majesty and provision. God's command to tend and keep the earth (Genesis 2) did not pass away with the fall of man; we are still responsible. Therefore, we humbly take responsibility for the damage we have done to God's cosmic revelation and pledge to take an unwavering stand to preserve and protect the creation over which we have been given responsibility by Almighty God Himself.

2. It Is Prudent to Address Global Climate Change

Even in the absence of perfect knowledge, we have to make informed decisions about the future. We do not believe unanimity is necessary for prudent action. We can make wise decisions even in the absence of infallible evidence. Though the claims of science are neither infallible nor unanimous, they are substantial and cannot be dismissed.... Guided by biblical principle, we resolve to engage this issue without further lingering.... Humans must take responsibility for our contributions to climate change....

3. Christian Moral Doctrines Demand Environmental Stewardship

While we cannot review all relevant Christian convictions, we emphasize the following:

> We must care about climate issues because of our love for God — "the Creator, Redeemer and Ruler of the Universe." This is not our world, it is God's.
> We must care about environmental issues because of our commitment to God's Holy and inerrant Word, which is "the supreme standard by which all human conduct, creeds and religious opinions should be tried." When God made mankind, He commissioned us to exercise stewardship over the earth and its creatures (Gen. 1:26-28). Our motivation ... is primarily biblical.
> We must care about climate issues because we are called to love our neighbors... and to care for the "least of these."

Love of God, love of neighbor and Scripture's stewardship demands provide enough reason for Southern Baptists and Christians everywhere to respond to these problems....

4. It Is Time for Individuals, Churches, Communities and Governments to Act

"Every Christian should seek to bring industry, government and society under the sway of the principles of righteousness, truth and brotherly love." We realize that simply affirming our God-given responsibility to care for the earth will produce no tangible results. Therefore, we pledge to find ways to curb ecological degradation by promoting biblical stewardship and increasing awareness in our homes, businesses and our local churches. Many of our churches do not actively preach, promote or practice biblical creation care. We urge churches to begin doing so. The primary impetus for prudent action must come from the people.... We pledge, therefore, to give serious consideration to responsible policies that address the conditions set forth in this declaration.

Conclusion

In accordance with our Christian moral convictions and Southern Baptist doctrines, we pledge to act on... this document. We will not only teach the truths communicated here but also seek ways to implement the actions that follow.... In the name of Jesus Christ our Lord, we urge all who read this declaration to join us in this effort. *Laus Deo!*

African American Clergy

An Open Letter on Climate Change

The Earth is the Lord's and everything in it; the world, and all who live in it.

Psalm 24:1

WE ARE WRITING TO SUPPORT AND JOIN OUR POLITICAL LEADERS in taking bold action to address climate change. As African-Americans and members of the Christian clergy, we speak from the dual perspectives of a people whose longstanding bond with the Earth is equaled by an abiding commitment to walking by faith—a faith that serves as our moral compass, and directs us to be responsible stewards of the whole of creation.

Our perspective is shaped by the life-giving and healing legacy of earth-connectedness passed down by our forbears. A powerful connection to the Earth—among both urban and rural communities—has helped sustain the bodies and souls of our people for generations. Issues related to climate change are already affecting us, and without decisive action to protect our planet and its inhabitants, the hopes we hold for future generations will not be realized.

We affirm the interconnectedness of all humanity, and the importance of respecting and enhancing the lives of each and every one of our sisters and brothers—whether in Africa, Antarctica, Asia, Australia, Europe, North America or South America. We all inhabit the same Earth.... breathe its air and drink its water. When natural resources are degraded or scarce, the impact on an individual, group, or population is cause for concern, but more than that, it requires corrective action, and whenever possible, preventive measures. We are well aware of this sad reality: The voices of communities whose inhabitants look like us often are dismissed or disregarded. But the world cannot afford to silence us, and we cannot afford to be—and will not—be silent.

Climate change most directly impacts the poor and marginalized, but ultimately, everyone is in jeopardy. We must come together to solve this crisis, which can only be addressed effectively when we seek the good of all and speak candidly about where we are and where we've come from.

We grieve the long, sad history of racial oppression and the legacy of slavery that continues to affect virtually every dimension of life in America. As we assess the state of our country—and world—during the second term of our country's first Black president, we cannot escape the reality that our community has been deeply wounded by the systemic racism that reverberates throughout American society, curtailing opportunity and injuring psyches. We are deeply saddened by ongoing disparities in health care, housing, and economic opportunities, and in the educational and judicial systems. The frequency with which the "I-was-afraid-so-I-had-to-shoot" defense is successfully invoked when people of color are murdered reflects the pervasive impulse of law enforcement (and others) to view us as threatening or menacing, even in the most benign circumstances.

We know the fear that comes from knowing that none of us can control another person's perceptions or reactions. And yet, we live in hope — born of faith, reinforced by the progress thus far, and propelled forward by people of good will, committed to constructive engagement around difficult, pressing issues. Confronting climate change is one of those pressing issues.

We recognize that the church has been instrumental in advancing racial and Environmental Justice. Our desire to further extend that involvement into the realm of Climate Justice leads us to speak about the importance of caring for the environment we all share, even as we seek better ways of coexisting.... The effects of climate change.... are painfully clear: Over 50,000 people died of heat waves in 2010 in Europe; drought and increased wildfires threaten food supplies and communities throughout the United States and the world; drinking water supplies have been jeopardized, and respiratory illnesses such as asthma are particularly vulnerable to the exacerbation of heat-related illness and stress. Devastating environmental, health, social, and economic consequences imperil our brothers and sisters around the world and in the United States.

Today, as we contemplate the gravity of the global climate crisis, we think especially of communities of color and the poor. We are aware of the disproportionate impact that climate change is taking on these populations.... We know it is crucial that faith communities join with government, the private sector, and nonprofits in mitigation and resiliency planning, as more disruptions and damage caused by massive storms and tornados will continue to take place. This is a humanitarian emergency that requires resources to help vulnerable communities prepare for the dangers ahead and to care for those in harm's way.

We pledge to address climate change in the social justice tradition of the African-American church, with both spiritual and practical involvement. We especially encourage our civic leaders to help the marginalized and disenfranchised cope with... climate change, and we pledge to work with the college and university systems to develop solutions. We recognize the "inescapable network of mutuality" that Rev. Martin Luther King, Jr. described when he wrote "Injustice any-where is a threat to justice everywhere....Whatever affects one directly, affects all indirectly." The effects of climate change are significant, and our environment itself is an "inescapable network of mutuality."

We call for bold action from political leaders... that will: (1) Assure resources for community resiliency and for faith organizations in the face of immediate climate related disasters. (2) Create, promote and enable energy and carbon emission reduction targets. (3) Develop local communication networks to explore solutions for climate challenges. (4) Continue support of grants and research to provide targeted assistance to communities impacted by climate change around the world.

The upcoming United Nations Framework Convention on Climate Change (UNFCCC) negotiations must result in a binding international treaty that:

- Commits signatories to meeting the IPCC-agreed emission reduction levels necessary to keep temperature increases below 2°C;

- Includes financial commitments from wealthy countries to assist developing countries... in identifying and implementing adaptation and mitigation strategies;

- Includes a meaningful role in negotiations for those who are most severely impacted by climate change.

We pray for wisdom and courage to attend our elected officials and all political leaders. It is vital that steps are taken now; the consequences of inaction are dire.

The Lord God took the man and put him in the
Garden of Eden to work it and take care of it.

- Genesis 2:15

God is still speaking,

UNITED CHURCH
O F C H R I S T

RESOLUTION ON CLIMATE CHANGE

2007 GS-26

WHEREAS, the impact of global warming, as currently predicted and understood by leading scientists and scientific bodies around the world in reports of the UN's Intergovernmental Panel on Climate Change, as well as in reports of the National Aeronautics and Space Administration and the National Academy of Sciences, will dramatically and negatively alter God's gracious gift of creation;

WHEREAS, the effects of global warming are already clearly evidenced in the melting of glaciers and shrinking of the polar caps, threatening the polar bear with extinction and the Native Peoples of the Artic with loss of food resources, land, ancient traditions and ways of being in the world;

The effects of global warming are already clearly evidenced....

WHEREAS, experts speak with a profound sense of urgency and clearly state that the window of opportunity to avoid catastrophic climate change is rapidly diminishing;

WHEREAS, the predicted impact of global warming will have a disproportionate impact on those living in poverty, least developed countries, the elderly and children and those least responsible for the emissions of greenhouse gases;

The predicted impact of global warming will have a disproportionate impact on those living in poverty, least developed countries, the elderly and children...

THEREFORE, BE IT RESOLVED that the Twenty-sixth General Synod of the United Church of Christ admits Christian complicity in the damage human beings have caused to the earth's climate system and other planetary life systems, and urges recommitment to the Christian vocation of responsible stewardship of God's creation, and expresses profound concern for the pending environmental, economic, and social tragedies threatened by global warming, to creation, human communities and traditional sacred spaces;

United Church of Christ

WE FURTHER RESOLVE that the Twenty-sixth General Synod of the United Church of Christ urges the United States Government to respond to global warming with great urgency and firm leadership by supporting mandatory measures that reduce the absolute amount of green house gas emissions, and in particular emissions of carbon dioxide, to levels recommended by nationally and internationally recognized and respected scientific bodies;

WE FURTHER RESOLVE that the Twenty-sixth Synod of the United Church of Christ urges state and local governments to support and invest in energy conservation and, specifically, in sustainable, renewable and affordable systems of transportation, and calls on business and industry to lead in responses to global warming through increased investments in efficient and sustainable energy technologies that are economically accessible and just;

The United Church of Christ urges the U.S. Government to respond to global warming with great urgency and firm leadership by supporting mandatory measures that reduce the absolute amount of green house gas emissions....

WE FURTHER RESOLVE that the Twenty-sixth General Synod of the United Church of Christ urges all segments of the Church to address global warming in their decisions and investments and in their educational and advocacy efforts;

TO THAT END, the Twenty-sixth General Synod of the United Church of Christ calls on the Covenanted Ministries of the United Church of Christ, specifically Wider Church Ministries and Justice and Witness Ministries, to address the severe nature of this global warming crisis as one of the most urgent threats to humankind and, indeed, all of God's precious planet earth, and that Local Church Ministries develop materials to help churches "green" their buildings.

The United Church of Christ calls on... Church Ministries... to address the severe nature of this global warming crisis as one of the most urgent threats to humankind....

FUNDING

Funding for the implementation of this resolution will be made in accordance with the overall mandates of the affected agencies and the funds available.

The Religious Society of Friends (Quakers)

Quaker Earthcare Witness: Seeking emerging insights into right relationship with Earth and unity with nature.

Facing the Challenge of Climate Change

A shared statement by Quaker groups

September, 2014

"It would go a long way to caution and direct people in their use of the world, that they were better studied and knowing in the Creation of it. For how could [they] find the confidence to abuse it, while they should see the great Creator stare them in the face, in all and every part of it?"

- William Penn, 1693

AS QUAKERS, WE ARE CALLED TO WORK FOR THE PEACEABLE KINGDOM OF GOD on the whole Earth, in right sharing with all peoples. We recognize a moral duty to cherish creation for future generations.

As we gather at events surrounding the UN Climate Summit, we call on our leaders to make the radical decisions needed to create a fair, sufficient and effective international climate change agreement.

As Quakers, we understand anthropogenic (due to human activities) climate change to be a symptom of a greater challenge: how to live sustainably and justly on this Earth.

We recognize that current and unprecedented rates of greenhouse gas emissions, if left unchecked, will likely lead to global mean temperature rises of extreme detriment to human beings.

We recognize that catastrophic anthropogenic climate change is not inevitable if we choose to act.

We recognize a personal and collective responsibility to ensure that the poorest and most vulnerable peoples now, and all our future generations, do not suffer as a consequence of our actions.

We see this as a call to conscience.

The Religious Society of Friends (Quakers)

We recognize the connections between climate change and global economic injustice as well as unprecedented levels of consumption, and question assumptions of unlimited economic growth on a planet with limited natural resources.

We recognize that most greenhouse gas emissions are created by fossil fuel combustion. We recognize that our increasing population continues to pursue fossil fuel-dependent economic growth.

We recognize that the Earth holds more fossil fuel reserves than are safe to burn, and that the vast majority of proven fossil fuel reserves must remain in the ground if we are to prevent the catastrophic consequences of climate change. We therefore question profoundly the continued investment in, and subsidizing of, fossil fuel extraction.

We seek to nurture a global human society that prioritizes the well-being of people over profit, and lives in right relationship with our Earth; a peaceful world with fulfilling employment, clean air and water, renewable energy, and healthy thriving communities and ecosystems.

This week, we join the People's Climate March as members of this beautiful human family, seeking meaningful commitments from our leaders and ourselves, to address climate change for our shared future, the Earth, and the generations to come.

We see this Earth as a stunning gift that supports life. It is our only home. Let us care for it together.

Quaker Earthcare Witness (QEW)
Quaker United Nations Office (QUNO)
Friends Committee on National Legislation (FCNL)
Friends World Committee for Consultation (FWCC)
Canadian Friends Service Committee (CFSC)
Quaker Council for European Affairs (QCEA)
American Friends Service Committee (AFSC)
Princeton Friends Meeting, New Jersey, USA
Westtown Monthly, Pennsylvania, USA
Woodbrooke Quaker Study Centre, UK
FWCC- Asia West Pacific Section
New York Yearly Meeting
Memphis Friends Meeting
EcoQuakers Ireland
Quakers in Britain
Living Witness

If you need more information, please contact Lindsey Cook
at email: lfcook@quno.ch

Resolution of the Central Conference of American Rabbis

CLIMATE CHANGE

Adopted by the 116th Annual Convention
of the Central Conference of American Rabbis
Houston, Texas
March, 2005, 2011

Background

In December 1997, the nations of the world gathered in Kyoto, Japan to develop a treaty with binding commitments to address the threat of climate change. The International Panel on Climate Change (IPCC), a group of over 2,000 climate scientists was charged to evaluate the data on climate change to inform the treaty negotiations. IPCC has documented changes in the earth's atmosphere that are attributed to human activity, causing elevated levels of carbon dioxide and other greenhouse gasses that are heating the earth's surface.

The following Jewish and secular moral principles serve as the foundation for the Conference's position on the development of agreements and policies to address climate change:

Responsibilities to Future Generations:

"Therefore choose life, that you and your descendants may live" (Deuteronomy 30:20). Humankind has a solemn obligation to improve the world for future generations. Minimizing climate change requires us to learn how to live within the Earth's ecological limits so that we will not compromise ecological or economic security of those who come after us.

Integrity of Creation:

"The human being was placed in the Garden of Eden to till it and to tend it." (Genesis 2:15). Humankind has a solemn obligation to protect the integrity of ecological systems, so that their diverse constituent species, including humans, can thrive.

Protection of the Vulnerable:

"When one loves righteousness and justice, the earth is full of the loving-kindness of the Eternal" (Psalm 33:5). The requirements and procedures to address climate change must protect those most vulnerable to climate change both in the United States and around the globe: poor people, those living in coastal areas, those who rely on subsistence agriculture.

Equitable Distribution of Responsibility:

Nations' responsibilities for reducing greenhouse gas emissions should correlate to their contribution to the problem. The United States has built an economy highly dependent upon fossil fuel use that has affected the entire globe, and must therefore reduce greenhouse gas emissions in a manner that corresponds to its share of the problem.

Sustainable Development:

The Earth cannot sustain the levels of resource exploitation currently maintained by the developed world. As we work toward global economic development, the developed world should promote renewable energy sources and new technologies, so that developing nations do not face the same environmental challenges that we face today because of industrialization.

Strong action to reduce greenhouse gas emissions is consistent with a number of long-standing public policy priorities, including: improving air quality, increasing mass transit, development of non-polluting alternative energy sources, energy efficiency and energy conservation.

Together, the people of the world must use our God-given gifts to develop innovative strategies to meet [present] needs without compromising the ability of future generations to meet their own needs.

THEREFORE the Central Conference of American Rabbis resolves to:

1. Call on the United States Congress to take leadership on the issue of Global Climate Change. [we must scale] back emissions to year 2000 levels....

2. Urge the federal government to immediately adopt policies to accomplish emissions reductions, including: programs that use pricing to lower demand for fossil fuels; encouraging non-polluting energy sources; and raising revenue for public projects such as mass transit, that would lower carbon emissions.

3. Urge the federal government to complement the above policies with programs to help those Americans whose economic security would be jeopardized by such policies, including assistance to poor people and retraining for coal miners and other affected workers;

4. Urge the federal government to work cooperatively with other nations to address climate change through participation in international bodies, treaties and protocols and through the promotion of international development efforts that promote environmental sustainability; and

5. Urge institutions within the Jewish community to promote and provide resources to conduct energy audits of private homes and communal facilities, including synagogues, schools, community centers and commercial buildings and to promote eco-friendly purchasing.

Resolution on the Environment: A Green Covenant

WHEREAS the Rabbinical Assembly... has articulated the Jewish responsibility to address global climate change and work towards an environmentally responsible energy policy on a personal, communal, and national level with 14 resolutions since 1991 on protecting and enhancing the environment and the quality of life for all God's creatures; and

WHEREAS the urgency of such action has been dramatically substantiated during this past year by the international scientific community (U.N. Intergovernmental Panel on Climate Change 2007); and

WHEREAS this call to action must have as its central emphasis the reduction of carbon emissions and the lowering of our carbon footprint.

THEREFORE BE IT RESOLVED that the Rabbinical Assembly calls on its members world-wide to adopt the following policy and provisions in each of our institutions:

THE GREEN COVENANT: A JEWISH PLEDGE TOWARD CARBON NEUTRALITY

- ◆ We will calculate our institution's carbon footprint, and devise a plan of energy conservation and the use of renewable energy, and/or carbon offsets with a goal of attempting to achieve a 50% reduction in carbon emissions within a five year period.

- ◆ We will educate and advocate for carbon neutrality in our institutions and to encourage and to help facilitate carbon reduction strategies in the homes of those in our communities. We support the current campaign of the Federation of Jewish Men's Clubs including the Solar Ner Tamid Project (www.fjmc.org).

- ◆ We will support legislative efforts at the local, state, national, and international levels that promote the shift from high-carbon to low-carbon energy production, such as CAFE (Corporate Average Fuel Economy) standards and public policy that promotes the use of renewable resources such as by exploring alternate sources of energy.

- ◆ We will undertake a special campaign to support habitat preservation and reforestation efforts worldwide and to encourage our constituents to plant appropriate trees at home and in Israel.

- ◆ We will examine our investment policies with the aim of enhancing our socially responsible portfolio to support businesses and communities that share our environmental goals.

- ◆ We will urge our institutions to reduce energy use and carbon output in transportation by using public transit and/or purchasing high mileage or hybrid vehicles.

Passed by the Rabbinical Assembly Plenum, February, 2008, Washington, DC.

A Pan-Buddhist Declaration on Global Climate Change

The Time to Act is Now

TODAY WE LIVE IN A TIME OF GREAT CRISIS, confronted by the gravest challenge that humanity has ever faced: the ecological consequences of our own collective karma. The scientific consensus is overwhelming: human activity is triggering environmental breakdown on a planetary scale. Global warming, in particular, is happening much faster than previously predicted, most obviously at the North Pole.... In 2007 the Inter-governmental Panel on Climate Change (IPCC) forecast that the Arctic might be free of summer sea ice by 2100. It is now apparent that this could occur within a decade or two. Greenland's vast ice-sheet is also melting more quickly than expected. The rise in sea-level this century will be at least one meter—enough to flood many coastal cities and vital rice-growing areas such as the Mekong Delta in Vietnam.

Glaciers all over the world are receding quickly. If current economic policies continue, the glaciers of the Tibetan Plateau, source of the great rivers that provide water for billions of people in Asia, are likely to disappear by mid-century. Severe drought and crop failures are already affecting Australia and Northern China. Major reports—from the IPCC, United Nations, European Union, and International Union for Conservation of Nature—agree that, without a collective change of direction, dwindling supplies of water, food and other resources could create famine conditions, resource battles, and mass migration by mid-century—perhaps by 2030, according to the U.K.'s chief scientific advisor.

Global warming plays a major role in other ecological crises, including the loss of many plant and animal species that share this Earth with us. Oceanographers report that half the carbon released by burning fossil fuels has been absorbed by the oceans, increasing their acidity by about 30%. Acidification is disrupting calcification of shells and coral reefs, as well as threatening plankton growth, the source of the food chain for most life in the sea.

Eminent biologists and U.N. reports concur that "business-as-usual" will drive half of all species on Earth to extinction within this century. Collectively, we are violating the first precept—"do not harm living beings." And we cannot foresee the biological consequences for human life when so many species that contribute to our own well-being vanish....

Many scientists have concluded that the survival of human civilization is at stake. We have reached a critical juncture in our biological and social evolution. There has never been a more important time in history to bring the resources of Buddhism to bear on behalf of all living beings. The four noble truths provide a framework for diagnosing our current situation and formulating appropriate guidelines—because the threats and disasters we face ultimately stem from the human mind, and therefore require profound changes within our minds. If personal suffering stems from craving and ignorance—from the three poisons of greed, ill will, and delusion—the same applies to the suffering that afflicts us on a collective scale. Our ecological emergency is a larger version of the perennial human predicament. Both as individuals and as a species, we suffer from a sense of self that feels disconnected not only from other people but from the Earth itself. As Thich Nhat Hanh has said, "We are here to awaken from the illusion of our separateness."

We need to wake up and realize that the Earth is our mother as well as our home—and in this case the umbilical cord binding us to her cannot be severed. When the Earth becomes sick, we become sick, because we are part of her.

Our present economic and technological relationships with the biosphere are unsustainable. To survive the rough transitions ahead, our lifestyles and expectations must change. This involves new habits as well as new values. The Buddhist teaching that the overall health of the individual and society depends upon inner well-being, and not merely upon economic indicators, helps us determine the personal and social changes we must make.

Individually, we must adopt behaviors that increase everyday ecological awareness and reduce our "carbon footprint." Those of us in the advanced economies need to retrofit and insulate our homes and workplaces for energy efficiency; lower thermostats in winter and raise them in summer; use high efficiency light bulbs and appliances; turn off unused electrical appliances; drive fuel-efficient cars possible, and reduce meat consumption in favor of a environmentally-friendly plant-based diet.

These personal activities will not by themselves be sufficient to avert calamity. We must also make institutional changes, both technological and economic. We must "de-carbonize" our energy systems quickly by replacing fossil fuels with renewable energy sources that are limitless, benign and harmonious.... We especially need to halt... new coal plants, since coal is the most polluting and most dangerous source of atmospheric carbon. Wisely utilized, wind power, solar power, tidal power, and geothermal power can provide all the electricity that we require without damaging the biosphere.... We must also reverse the destruction of forests, especially the vital rainforest belt where most species of plants and animals live.

It has recently become quite obvious that significant changes are also needed in the way our economic system is structured. Global warming is intimately related to the gargantuan quantities of energy that our industries devour to provide the consumption levels that many of us have learned to expect. From a Buddhist perspective, a sane and sustainable economy would be governed by the principle of sufficiency: the key to happiness is contentment rather than an ever-increasing abundance of goods. The compulsion to consume more and more is an expression of craving, the very thing the Buddha pinpointed as the root of suffering.

Instead of an economy that emphasizes profit and requires perpetual growth to avoid collapse, we need to move together towards an economy that provides a satisfactory standard of living for everyone while allowing us to develop our full potential (including spiritual) in harmony with the biosphere that sustains and nurtures all beings.... If political leaders are unable to recognize the urgency of our global crisis, or unwilling to put the long-term good of humankind above the short-term benefit of fossil-fuel corporations, we may need to challenge them with sustained campaigns of citizen action.

Dr James Hansen and other climatologists have recently defined the precise targets needed to prevent global warming from reaching catastrophic "tipping points." For human civilization to be sustainable, the safe level of carbon dioxide in the atmosphere is no more than 350 parts per million (ppm). This target has been endorsed by the Dalai Lama, along with Nobel laureates and distinguished scientists. Our current situation is worrisome in that the present level is already 387 ppm, and has been rising at 2 ppm per year. We are challenged not only to reduce carbon emissions, but also to remove large quantities of carbon gas already present in the atmosphere.

As signatories to this statement of Buddhist principles, we acknowledge the urgent challenge of climate change. We join with the Dalai Lama in endorsing the 350 ppm target. In accordance with Buddhist teachings, we accept our individual and collective responsibility to do whatever we can to meet this target, including... the personal and social responses outlined above.

We have a brief window of opportunity to take action, to preserve humanity from imminent disaster and to assist the survival of the many diverse and beautiful forms of life on Earth. Future generations, and other species that share the biosphere with us, have no voice to ask for our compassion, wisdom, and leadership. We must listen to their silence. We must be their voice, too, and act on their behalf.

Hindu Declaration on Climate Change

Statement at the Convocation of Hindu Spiritual Leaders
Parliament of the World's Religions
Melbourne, Australia, December 8, 2009

Earth, in which the seas, the rivers and many waters lie, from which arise foods and fields of grain, abode to all that breathes and moves, may She confer on us Her finest yield.

- Bhumi Suktam, Atharva Veda xii.1.3

THE HINDU TRADITION UNDERSTANDS that man is not separate from nature, that we are linked by spiritual, psychological and physical bonds with the elements around us. Knowing that the Divine is present everywhere and in all things, Hindus strive to do no harm. We hold a reverence for life and an awareness that the great forces of nature —the earth, water, fire, air and space—as well as all the various orders of life, plants and trees, forests and animals, are bound to each other in life's cosmic web.

Our beloved Earth, so touchingly looked upon as the Universal Mother, has nurtured mankind through millions of years…. Now centuries of rapacious exploitation of the planet have caught up with us, and a radical change in our relationship with nature is no longer an option. It is a matter of survival. We cannot continue to destroy nature without destroying ourselves. The problems besetting our world—war, disease, poverty and hunger—will all be magnified by the predicted impacts of climate change.

The nations of the world have yet to agree upon a plan to ameliorate man's contribution to this complex change. This is largely due to powerful forces in some nations which oppose any such attempt, challenging the very concept that unnatural climate change is occurring. Hindus everywhere should work toward an international consensus. Humanity's very survival depends upon our capacity to make a major transition of consciousness, equal in significance to earlier transitions from nomadic to agricultural, agricultural to industrial and industrial to technological. We must transit to complementarity in place of competition, convergence in place of conflict, holism in place of hedonism…. We must, in short, move rapidly toward a global consciousness that replaces the present fractured and fragmented consciousness of the human race.

Mahatma Gandhi urged, "You must be the change you wish to see in the world." If alive today, he would call upon Hindus to set the example, to change our lifestyle, to simplify our needs and restrain our desires. As one sixth of the human family, Hindus can have a tremendous impact. We can and should take the lead in Earth-friendly living, personal frugality, lower power consumption, alternative energy, sustainable food production and vegetarianism, as well as in technologies that address our shared plight…. Thus, in the spirit of *vasudhaiva kutumbakam*, "the whole world is one family," Hindus encourage the world to be prepared to respond with compassion to such calamitous challenges as population displacement, food and water shortage, catastrophic weather and rampant disease.

Hindus still know we must do all that is humanly possible to protect the Earth and her resources for the present as well as future generations.

The Anchorage Declaration

The following Declaration is from the Indigenous Peoples'
"Global Summit on Climate Change,"
Anchorage, Alaska
April 20 – 24th, 2009

Introduction

From April 20-24, 2009, representatives of Indigenous peoples from the Arctic, North America, Asia, Pacific, Latin America, Africa, Caribbean and Russia met in Anchorage, Alaska for the Indigenous Peoples' Global Summit on Climate Change. We thank the Ahtna and the Dena"ina Athabascan Peoples in whose lands we gathered.

WE EXPRESS OUR SOLIDARITY AS INDIGENOUS PEOPLES living in areas that are the vulnerable to the impacts and causes of climate change. We reaffirm the unbreakable and sacred connection between land, air, water, oceans, forests, sea ice, plants, animals and our human communities as the material and spiritual basis for our existence.

We are deeply alarmed by the accelerating climate devastation brought about by unsustainable development. We are experiencing profound and disproportionate adverse impacts on our cultures, human and environmental health, human rights, well-being, traditional livelihoods, food systems and food sovereignty, local infrastructure, economic viability, and our very survival as Indigenous Peoples.

Mother Earth is no longer in a period of climate change, but in climate crisis. We therefore insist on an immediate end to the destruction and desecration of the elements of life.

Through our knowledge, spirituality, sciences, practices, experiences and relationships with our traditional lands, territories, waters, air, forests, oceans, sea ice, other natural resources and all life, Indigenous Peoples have a vital role in defending and healing Mother Earth. The future of Indigenous Peoples lies in the wisdom of our elders, the restoration of the sacred position of women, the youth of today and in the generations of tomorrow.

We uphold that the inherent and fundamental human rights and status of Indigenous Peoples, affirmed in the United Nations Declaration on the Rights of

Indigenous Peoples (UNDRIP), must be fully recognized and respected in all decision-making processes and activities related to climate change. This includes our rights to our lands, territories, environment and natural resources as contained in Articles 25—30 of the UNDRIP.

When specific programs and projects affect our lands, territories, and natural resources, the right of self-determination of Indigenous Peoples must be recognized and respected, emphasizing our right to Free, Prior and Informed Consent, including the right to say "no."

Calls for Action:

1. In order to achieve the fundamental objective of the United Nations Framework on Climate Change (UNFCCC), we call upon the fifteenth meeting of the Conference of the Parties (COP-15) ... to support a binding emissions reduction target for developed countries of at least 45% below 1990 levels by 2020 and at least 95% by 2050. In recognizing the root causes of climate change, we call upon States to work towards decreasing dependency on fossil fuels. We further call for a just transition to decentralized renewable energy economies, sources and systems owned and controlled by our local communities to achieve energy security and sovereignty.

In addition, Summit participants agreed to present two options for action which were each supported by one or more of the participating regional caucuses. These were as follows:

> A. We call for the phase out of fossil fuel development and a moratorium on new fossil fuel developments on or near Indigenous lands and territories.

> B. We call for a process that works towards the eventual phase out of fossil fuels, without infringing on the right to development of Indigenous nations.

2. We call upon the UN to recognize the importance of our Traditional Knowledge and practices shared by Indigenous Peoples in developing strategies to address climate change. To address climate change we also call on the UNFCCC to recognize the historical and ecological debt of the Annex 1 countries in contributing to greenhouse gas emissions. We call on these countries to pay this historical debt.

3. We call on the Intergovernmental Panel on Climate Change (IPCC), and other relevant institutions to support Indigenous Peoples in carrying out Indigenous Peoples' climate change assessments.

4. We call upon the UNFCCC's decision-making bodies to establish formal structures and mechanisms for and with the full and effective participation of Indigenous Peoples. Specifically we recommend that the UNFCCC:

> a. Organize regular Technical Briefings by Indigenous Peoples on Traditional Knowledge and climate change;

b. Recognize and engage the International Indigenous Peoples' Forum on Climate Change and its regional focal points in an advisory role;

c. Immediately establish an Indigenous focal point in the secretariat of the UNFCCC;

d. Appoint Indigenous Peoples' representatives in UNFCCC funding mechanisms in consultation with Indigenous Peoples;

e. Take the necessary measures to ensure the full and effective participation of Indigenous communities in formulating, implementing, and monitoring activities, mitigation, and adaptation relating to impacts of climate change.

5. All initiatives under Reducing Emissions from Deforestation and Degradation (REDD) must secure the recognition and implementation of the human rights of Indigenous Peoples, including security of land tenure, ownership, recognition of land title according to traditional ways, uses and customary laws and the benefits of forests for climate, ecosystems, and Peoples before taking any action.

Mother Earth is no longer in a period of climate change, but in climate crisis.

6. We challenge States to abandon false solutions to climate change that negatively impact Indigenous Peoples' rights, lands, air, oceans, forests, territories and waters. These include nuclear energy, large-scale dams, geo-engineering techniques, "clean coal," agro-fuels, plantations, and market based mechanisms such as carbon trading, the Clean Development Mechanism, and forest offsets. The human rights of Indigenous Peoples to protect our forests and forest livelihoods must be recognized, respected and ensured.

7. We call for adequate and direct funding in developed States and for a fund to enable Indigenous Peoples' full and effective participation in all climate processes, including adaptation, mitigation, monitoring and transfer of appropriate technologies in order to foster our empowerment, capacity-building, and education.

We urge relevant United Nations bodies to facilitate and fund the participation, education, and capacity building of Indigenous youth and women to ensure engagement in all international and national processes related to climate change.

8. We call on financial institutions to provide risk insurance for Indigenous Peoples to allow them to recover from extreme weather events.

9. We call upon all United Nations agencies to address climate change impacts in their strategies and action plans, in particular their impacts on Indigenous Peoples, including the World Health Organization (WHO), United Nations Educational, Scientific and Cultural Organization (UNESCO), and United Nations Permanent Forum on Indigenous Issues (UNPFII).

10. We call on the United Nations Environment Programme (UNEP) to conduct an assessment of short-term drivers of climate change, specifically black carbon, with a view to initiating negotiation of an agreement to reduce emission of black carbon.

11. We call on States to recognize, respect and implement the human rights of Indigenous Peoples, including rights to traditional ownership, use, access, occupancy and title to traditional lands, air, forests, waters, oceans, sea ice and sacred sites as well as to ensure that the rights affirmed in Treaties are upheld and recognized in land use planning and climate change mitigation strategies. In particular, States must ensure that Indigenous Peoples have the right to mobility and are not forcibly removed or settled away from their traditional lands and territories, and that the rights of Peoples in voluntary isolation are upheld. In the case of climate change migrants, appropriate programs and measures must address their rights, status, conditions, and vulnerabilities.

12. We call upon states to return and restore lands, territories, waters, forests, oceans, sea ice and sacred sites that have been taken from Indigenous Peoples, limiting our access to our traditional ways of living, thereby causing us to misuse and expose our lands to activities and conditions that contribute to climate change.

13. In order to provide the resources necessary for our collective survival in response to the climate crisis, we declare our communities, waters, air, forests, oceans, traditional lands and territories to be *"Food Sovereignty Areas,"* defined and directed by Indigenous Peoples according to customary laws, free from extractive industries, deforestation and chemical-based industrial food production systems (i.e. contaminants, agro-fuels and genetically modified organisms).

14. We encourage our communities to exchange information while ensuring protection and recognition of and respect for the intellectual property rights of Indigenous Peoples at the local, national and international levels pertaining to our Traditional Knowledge, innovations, and practices. These include knowledge and use of land, water and sea ice, traditional agriculture, forest management, ancestral seeds, pastoralism, food plants, animals and medicines and are essential in developing climate change adaptation and mitigation strategies, restoring our food sovereignty and food independence, and strengthening our Indigenous families and nations.

Further, We offer to share with humanity our Traditional Knowledge, innovations, and practices relevant to climate change, provided our fundamental rights as inter-generational guardians of this knowledge are recognized and respected. We reiterate the urgent need for collective action.

*Agreed by consensus of all the participants in
the Indigenous Peoples' Global Summit on Climate Change*

*Anchorage Alaska
April 24th 2009*

Islamic Foundation for Ecology & Environmental Sciences

المؤسسة الإسلامية للطبيعة وعلوم البيئة

A Muslim Guide to Climate Change

As for the earth, We have spread it out, set firm mountains in it, and made everything grow there in due balance.

- Qur'an 15:19 H

Those who squander are the brothers of Satan, and Satan is most ungrateful to his Lord.

- Qur'an 17:27 H

The state of planet earth, our only home, is rapidly deteriorating. Authoritative reports follow in rapid succession, warning of the dire consequences of climate change and global warming caused by human activity. What is now urgently needed is a dramatic change in the way the human race conducts its affairs and Muslims, who constitute one fifth of the world's population, can make a big difference if only they take these warnings seriously and heed the teachings of their faith.

What is Climate Change?

"Corruption has flourished on land and sea as a result of people's actions and He will make them taste the consequences of some of their own actions so that they may turn back" (30:41 H).

The Earth's climate (weather) changes all the time, but most scientists say that it is now starting to change suddenly, drastically and unpredictably, in ways that may well result in catastrophic floods, droughts, storms, forest fires, heat-waves, etc. We are already seeing an increase in the rate of such natural disasters around the world.

Many scientists agree that the most likely reason for these changes is a build-up of carbon dioxide in Earth's atmosphere, due to human beings' excessive burning of fossil fuels, such as oil, gas, and coal. This has increased the temperature of the earth by trapping heat from the sun (the 'Greenhouse Effect'), causing 'Global warming'.

This is being seen in many places but in particular in the polar ice caps and glaciers, which are melting at an unprecedented rate. As the ice melts, another 'greenhouse' gas, methane, is released from trapped bubbles and from the peat bogs that have been frozen for many centuries. Methane is highly flammable and contributes even more to global warming than carbon dioxide.

As the ice melts, the rate of global warming increases even faster, as the ice no longer reflects the heat of the sun out into space. Since human beings appear to have played an important part in upsetting the balance of the climate, people all over the world are being asked to find ways to cut their carbon emissions by abandoning fossil fuels and turning to renewable energy sources.

Why is it important for Muslims to take action?

Islam teaches Muslims to respect Allah's creation and maintain the balance He created. Allah commands people, *"Do not cause corruption on the earth"* (2:11).

"Climate change is a far greater threat to the world than international terrorism"

- UK Government chief scientific adviser Sir David King

While most Muslim countries are poor and use far less fossil fuels than rich ones, many Muslims also live in the rich countries or lead wasteful lifestyles and need to play their part in reducing their consumption of fossil fuels.

Many Muslim countries produce oil and so have a vested interest in maintaining the fossil fuel industry. Nevertheless, it's in their interest to keep the price of oil high, to maximise their income from production, and to make the oil last longer before it is all used up.

Higher oil prices would create incentives for people to switch to cleaner and more sustainable energy. Oil producing countries should therefore include sustainable energy industries in their long-term investment and diversification plans.

Authoritative reports follow in rapid succession,
warning of the dire consequences of climate change
and global warming caused by human activity.

LETTER FROM THE ISLAMIC FOUNDATION FOR ECOLOGY AND ENVIRONMENTAL SCIENCES

"In the name of Allah, the most kind, the most merciful. All praise is for Allah, and peace and blessings upon his Prophet Mohammad (saw) and all of his followers.

"Surely brothers and sisters, in our moments of reflection, as we gaze around at our Lord's signs and admire his majesty in the beauty of the trees that provide us with delicious fruits, the clouds from which falls the nourishing rain and the vast oceans which man harvests and upon which he is carried great distances; we must surely recognise that the earth and everything in it is a trust placed upon us. We are temporary custodians and this is another of Allah's tests for us to see how we care for our environment and all things within it.

"Climate change is the consequence of Man disturbing the balance of Allah's creation. The harmony of our weather, our land, air and sea has all been affected by too much pollution and too little care in consuming the earth's resources. It is required within our Islamic faith, that we behave as responsible and considerate citizens in our families and communities, practising concern and moderation in our lives.

"The state of planet earth, our only home, is rapidly deteriorating. Authoritative reports follow in rapid succession, warning of the dire consequences of climate change and global warming caused by human activity.

"What is now urgently needed is a dramatic change in the way the human race conducts its affairs and Muslims, who constitute one fifth of the world's population, can make a big difference if only they take these warnings seriously and heed the teachings of their faith."

- Hajj Fazlun Khalid, Founder Director
Islamic Foundation for Ecology and Environmental Sciences

World Council of Churches

A worldwide fellowship of 349 churches seeking unity,
a common witness and Christian service

Global Warming and Climate Change

"Be Stewards of God's Creation!"

*"In the beginning God created the heavens and the earth...
God saw all that he had made, and it was very good."*
(Genesis 1:1, 31, NIV)

The present statement builds on previous statements of the WCC,
adopted by the WCC executive committee.

September, 2007
abridged

1. The scriptures affirm that the "earth is the Lord's and everything in it" (Psalm 24:1). In Genesis 1:28, God charges humanity to care for the earth by giving humanity "dominion" over it. The word "dominion" is most appropriately translated as "stewardship," since humanity is not the master of the earth but steward to care for the integrity of creation. God wondrously and lovingly created a world with more than enough resources to sustain... human beings and living creatures. But humanity is not always faithful in its stewardship. Mindless production and excessive consumption... have led to continuous desecration of creation, including global warming....

3. Climate change, as the variation in the earth's global climate ... over time, and its effects are being experienced already in many regions of the world. Global warming... is one of the most evident aspects of climate change. The average temperature of the earth is rising.... Climate change raises ecological, social, economic, political and ethical issues, and demonstrates the brokenness of relationships between God, humankind and creation.

4. As stated by the "IPCC Report" and other studies, the situation needs urgent mitigation and adaptation measures to prevent further adverse consequences of rising temperatures. Mitigation (dealing with the causes) is a must for developed countries that will have to reduce their carbon dioxide (CO_2) emissions. Adaptation (dealing with the impacts) is urgently needed by developing countries to be able to cope with the changes that are happening.

5. To address the threats the world is facing because of climate change, action must be taken now. If there is no profound change in life styles, development patterns and the pursuit of economic growth, humanity will not meet the challenge. As the WCC delegation in Bali clearly stressed, "It is our conviction as members of faith communities that a 'change of paradigm' from one way of thinking to another is needed if we are to adequately respond to the challenge of climate change."

6. Climate change is both an environmental issue and a matter of justice. Major greenhouse gas (GHG) emitters have a historic responsibility to assume, to stop and to reverse the current trend. Developing countries, while seeking better conditions, face a dilemma which should be confronted in looking for ways not to repeat the path that led to the present situation. The current unsustainable production and consumption patterns have ... generated... "an ecological debt" towards humanity and the earth. To reverse this trend it becomes crucial to look for technologies and practices both to mitigate and adapt... to the needs of vulnerable communities.

7. Churches and religious communities can take leadership roles in addressing global warming and climate change.... The question... is whether we can rise together to meet this unprecedented opportunity. Churches and religious communities must find ways to challenge and motivate each other to measure our ecological and economic "footprints" and to make lasting changes in lifestyles and economic pursuits. Church members have to take responsibility for paying their share of the ecological debt that looms large in the years ahead.

8. As the effects of global warming can lead to conflict between populations competing over scarce resources, WCC member churches' actions with regards to climate change should be seen in relationship with the "Decade to Overcome Violence."

9. Many churches and ministries have started to take action concerning climate change and global warming. The Ecumenical Patriarch has played a leadership role advocating for the care of creation, and calling, on 1 September 1989, to observe September 1st... as creation day. This call was reiterated by the Third European Ecumenical Assembly, meeting in Romania in September 2007.

The World Council of Churches

Meeting in Geneva, Switzerland
13-20 February 2008

A. Urgently calls churches to strengthen their moral stand in relationship to global warming and climate change, recalling its adverse effects on poor and vulnerable communities.... and encourages churches to reinforce their advocacy in response to global warming and climate change;

B. Calls for a profound change in the relationship towards nature, economic policies, consumption, production and technological patterns.

D. Urges member churches to observe through prayers and action a special time for creation, its care and stewardship, starting on September 1st every year....

E. Requests theological schools, seminaries and academies to teach stewardship of all creation in order to deepen the ethical and theological understanding of the causes of global warming....

F. Promotes the exploration of inter-religious cooperation and constructive response, ensuring a better stewardship of creation and a common witness through concrete actions.

Coming to Terms with our Moral Responsibility

Climate Change and the Future of Civilization

July 7, 2014
abridged for space

OUR CIVILIZATION NOW FACES AN UNAVOIDABLE OPTION: Will we choose the difficult path that leads to hope? Or will we cater to wistful, hear-no-evil dreams from which we'll awaken into a bleak dystopia? The difficult but rewarding path brings us to a richer vision of ourselves, our Creator, and creation. We'll either re-forge the bonds between morality and policy or our dreams for the future will abandon us on the rubble of a modern-day Tower of Babel.

Such a choice compels the National Religious Coalition on Creation Care to call upon U.S. policymakers to heed God's instruction in the Book of Joshua: *"Be strong and very courageous"* (Joshua 1:7). Possess the courage to walk the path of hope. Be guided by the truth and formulate policies on that basis. Make the -right choices. Denying human-induced climate change may be easy in the short term, but it is profoundly immoral.

We urge courage upon America's sleeping giant – the 80% claiming religious affiliation – to awaken to their own religious teachings on climate change and their responsibilities as citizens and demand that political leaders act on a scale consonant with the evidence revealed by the scientific community. For the United States bears a special burden for leading the world on this issue because it is responsible for 26% of the greenhouse gas CO_2 that is in the atmosphere.

In the late 1970's, we received multiple warning from major scientists and the National Academy of Sciences that global warming would happen if we kept burning fossil fuels and that this could lead to a loss of our coastal cities. By the early and mid-1990's the evidence indicated that the planet was already warming and human beings were the primary drivers of this warming. The evidence was so clear that in the first years of this century military officials warned of potential devastation and consequent insecurity.

Religious leaders soon began issuing their appeals:

"Climate change is much more than an issue of environmental protection," said HAH Ecumenical Patriarch Bartholomew of Constantinople, adding, "insofar as it is human-induced, it is a profoundly moral and spiritual problem."

Pope Francis recently said: "Safeguard Creation, because if we destroy Creation, Creation will destroy us! Never forget this."

The Central Conference of American Rabbis said, "Together, the people of the world can, and must, use our God-given gifts to develop innovative strategies to meet the needs of all who currently dwell on this planet, without compromising the ability of future generations to meet their own needs." The rabbis called for "strong action" to reduce greenhouse gas emissions.

"The need to act now is urgent," assert dozens of Evangelical Christian leaders.

And Buddhist leaders jointly proclaimed: "The scientific consensus is overwhelming: human activity is triggering environmental breakdown on a planetary scale."

Yet our decision-makers repeatedly balked, potentially paving devastation's path. From 1951-1980, extremely hot summers occurred over less than 1% of Earth's surface, now, because of global warming, they occur over 10% of the land area. This shift is responsible for over 130,000 deaths in Europe in the first decade of the 21st Century. Extrapolations of this data indicate that by mid-century, when children born today are in their late thirties, the decadal death toll from temperature alone could be comparable to the number of deaths suffered during WW II.

Two independent scientific papers published in early May show unnerving evidence that six glaciers on the West Antarctica ice sheet are now in a phase of unstoppable melt that will lead to a sea level rise of 1.2 meters (4 feet). Eric Rignot, a NASA scientist and co-author of one of the studies, said the melt will "likely trigger the collapse of the rest of the West Antarctic ice sheet, which comes with a sea level rise of between three and five meters [10 to 16 feet]." It is uncertain how long this collapse will take, but a conservative estimate is several centuries. This is in addition to the increased melting of the Greenland ice sheet, which, according to another recent study, will contribute much more to sea level rise by 2100 than previously thought.

A three-foot rise would ruin Charleston, New Orleans, Fort Lauderdale, Tampa, Saint Petersburg, and Miami; a sixteen-foot rise would condemn Boston and Houston and dwindle San Diego, Seattle, and New York to remnants of their former selves – and those estimates do not factor in storm surge damage, which is likely to ravage our coasts much earlier. One study suggests that a five-foot sea level rise would bring Superstorm Sandy-style surges every other year along the East Coast, putting our cities under siege.

On our current path, over 1/3 of all plant and animal species, excluding ocean species, could be "committed to extinction" by mid-century.[v] By the mid 2060's carbon dioxide levels will be high enough to set in motion a mass extinction of ocean ecosystems.

This could bring about the Earth's 6th Mass Extinction. Humans cannot cope with the national and international disturbances that will occur as we destroy the food chains upon which we all depend.

Given the evidence, our responsibility, and the call, it is not an overstatement to label the continual denial of the evidence by some politicians and the feckless foot dragging by others as radically evil. Either we demand that our leaders act in accord with the truth of our situation and come together in a World War II-style effort to transform our economy to run on abundant renewable energy or we will abandon our own children to pick through Babel's debris in a world of mega-droughts, floods, wild fires, and endless wars over diminishing water and food supplies.

The ancient wisdom found in our sacred texts has revealed to us that we were never meant to conquer creation, but we are called to nurture God's earth. The conquering of creation, as is becoming all too evident, means our destruction. We are faced now with the starkest of choices as laid out by Moses in the Book of Deuteronomy:

> "*I have set before you life and death, blessings and curses. Choose life so that you and your descendants may live*" (Deuteronomy 30:19-20).

The Interreligious Ecojustice Network of Connecticut

THE HARTFORD DECLARATION

A Call for Action on Climate Change by Connecticut's Interfaith Community

November 7, 2013

- There comes a time in every generation when a matter of great urgency requires that we, who belong to Connecticut's diverse faith community, express our concerns with moral clarity and with a unified voice. That pivotal moment has arrived. We can no longer ignore the plain facts of climate change.

- Earth is increasingly under threat from climate change and global warming, endangering human beings and other life-forms in all regions of the globe. As a result, recent storms, floods, droughts, wildfires and heat waves have begun to have serious impact on our lives. Continuing increases of these extreme events threatens to destroy the underlying basis of human civilization as we know it.

  - Climate change is undermining the very ecological fabric of Earth that sustains life, while eliminating vast number of species on Earth.

  - Climate change disproportionately impacts society's most vulnerable members – the poor and underserved community, the elderly, the chronically ill, communities of colors, infants and young children, and those least able to fend for themselves.

  - Many developing regions of the world continue to be impacted by severe storms, drought-like conditions, sea level rise, vector-borne diseases, and loss of food security and safe drinking water. This is already causing mounting tension, conflict and war. As conditions worsen, it will cause millions of inhabitants to become environmental refugees with untold suffering, affecting the security and well-being of the entire globe.

- As members of the faith community, we have a deep obligation to understand the full dimensions of this growing problem, which the overwhelming consensus of the scientific community has documented in the past few decades.

- Safeguarding all creation on Earth is a sacred trust that is placed upon us – to love, to care for and to nurture. *We accept this trust* as a universal moral imperative, one that we share across all human societies, religious faiths and cultural traditions.

- Given the urgency of the current situation, *we solemnly pledge* to:

  - Foster a reflective and prayerful response to the threat of global climate change.

  - Work together as people of many religions and cultures to live sustainably on planet Earth.

  - Encourage members of our faith to develop and implement energy conservation plans and to use safe, clean, renewable energy.

  - Be an authentic witness for action on climate change and environmental justice through teaching, preaching and by letting our voices be heard in the public sphere.

  - Advocate for local, state, national and international policies and regulations that enable a swift transition from dependence on fossil fuels to safe, clean, renewable energy.

Faith Leaders Speak Out on the Need to Address Climate Change

AS RELIGIOUS GROUPS STUDIED THE CLIMATE PROBLEM, FAITH LEADERS awakened and became active across the world's faith traditions. Listen closely to these servants of God and moral teachers of humanity, as their words provide reflection and guidance to see deeper into the climate challenge.

Pope Francis has said: "Safeguard Creation, because if we destroy Creation, Creation will destroy us! Never forget this."

The Central Conference of American Rabbis in one voice has declared, "Together, the people of the world can, and must, use our God-given gifts to develop innovative strategies to meet the needs of all who dwell on this planet, without compromising future generations' ability to meet their own needs." The rabbis call for "strong action" to reduce greenhouse gas emissions.

"The need to act now is urgent," declare many Evangelical Christian leaders. "In the name of Jesus Christ our Lord, we urge all who read this [our] declaration to join us in this effort."

"Climate change is much more than an environmental protection issue," writes Ecumenical Patriarch Bartholomew of Constantinople. "Insofar as it is human-induced, it is a profoundly moral and spiritual problem."

Buddhist leaders have jointly proclaimed: "The scientific consensus is overwhelming: human activity is triggering environmental breakdown on a planetary scale."

Despite these strong and clear statements, some people persist in doubting the urgency of climate change. It is time for religious institutions to follow their leaders and teach the urgency of this emerging calamity. Failure to respond will result in unnecessary deaths and suffering because some could not discern the handwriting already writ boldly across the face of creation.

The Great Warming: ... Its More Than Just the Heat

An Interview with Reverend Richard Cizik
President of the New Evangelical Alliance for the Common Good

Q: Where is this evangelical concern for the environment coming from?

Rev. Cizik:

I would say that this newfound passion, this concern for Creation Care as we call it, comes straight from God and the Holy Spirit who is regenerating people's hearts to realize the imperative of the scriptures to care for God's world in new ways. It comes from God Himself. He has changed my heart too. I have had a conversion to this cause.

The climate change crisis that is occurring is not something we can wait ten years, five years, even a year, to address. Climate change is real and human induced. It calls for action soon. And we are saying action based upon a biblical view of the world as God's world. And to deplete our resources, to harm our world by environmental degradation, is an offense against God. That's what the Scriptures say. Therefore, if we are to be obedient to the Scriptures, there is no time to wait, no time to stall, no time to deliberate.

It's hard to know how many evangelicals have come to realize their full biblical responsibility. Surveys show that in the last election 52% of evangelicals believed that the unborn sanctity of human life was a priority, and yet, amazingly enough, 48% said that a clean environment was an important priority. So what explains the fact that they know this is important, yet they have largely sat on their hands? There are a lot of explanations. One is that environmentalism has a "left wing tilt" in their minds. And they haven't had pastors who preached on the importance of Creation Care. Most have not heard one sermon in their entire biblical life on this topic, if you can believe that.

I think what's occurring is that people in the pews have not been getting cues from their pastors and denominational leaders - the people they esteem most. But we know that once they hear this word from their spiritual heads, from their churches, denominations and religious organizations, they will respond. And they've told us they will respond. Not just by changing their lifestyle, but by changing their politicians. Now that's where the rubber meets the road.

Evangelicals comprise between 40-50% per cent of the Republican base, and it has been the leadership of the Republican party, sad to say, that has not shown the leadership it should on environmental issues… in fact, they have stymied action on climate change. So if the largest single population group in the Republican coalition were to say *"This is important, we want you, as our leaders in the Republican party, to take leadership on climate change, on the stewardship of our natural resources,"* if evangelical Christians were to say that, I dare say Republicans will listen.

As for the lack of action, the disconnect between smart people in this city – Washington, DC – the disconnect between the recognition that there is an obvious problem and the willingness to adopt an obvious solution, is explainable only by the fact that there are vested interests, political interests, who lobby against environmental action. Second, there is an ideological predisposition against regulation. And third, simple inertia. But the first cannot be dismissed, and there are oil and gas interests who have a reason not to want to take action on climate change.... They have a cause which I don't believe is in the interest of God, or in the interest of man. And I say that sadly.

Climate change is real and human induced. It calls for action soon.

Q: Aren't people in the church going to be uncomfortable working with environmental types?

Rev. Cizik:
Look, evangelical Christians have learned that we have to do politics differently than we used to. In the 1980's many an evangelical thought *"well, we'll simply convert the opposition to our side."* We've learned that we can't simply convert all those in public life to our side, especially on religion. But we do know that we can convert people to a common good that everyone can agree upon. For example, evangelical Christians collaborated with Tibetan Buddhists to pass religious freedom bills. They've collaborated with gay rights activists to pass the Global AIDS Initiative. So, can we not collaborate with environmentalists? Of course we can! But do we need to develop our own voice first? Sure, we need to. Evangelicals need to sense that they are speaking out of their own tradition, their own Biblical tradition – that they are not "me too" environmentalists. So, once evangelical Christians have developed their voice, once they are comfortable with the fact that there needs to be government action, and that voluntary action won't suffice for climate change, they will conclude that yes, environmentalism is not a bad word.

Q: Has something changed? Not all evangelicals are supportive of the Creation Care idea.

Rev. Cizik:
There is a debate going on within evangelical circles as to what is the highest priority. Is it to care for human life, unborn human life, first? It is to care about the poor? And how much place should we give to Creation Care, concern for environmentalism? I believe that the inevitable conclusion will be that all the issues are important and they're inter-related. For example, mercury pollution from coal burning utility plants falls into our rivers and is consumed by fish. This in turn impacts our children, because now one out of six women in America has unduly high levels of mercury in their systems - impacting unborn children. It's a sanctity of human life issue that relates to the environment.

Q: On a personal note, how did you get "converted" to environmentalism?

Rev. Cizik:
I've worked in Washington for twenty five years. Then in 2002, Jim Ball, a friend of mine who is a

leader in the evangelical environmental movement, dragged me to a conference at Oxford, England on this subject. And I heard from the evangelical Christian scientist Sir John Houghton – one of the drafters of the Intergovernmental Panel on Climate Change document – arguably one of the most influential scientists in the world, make the case that I couldn't shirk, shrug, rationalize or escape my Biblical responsibility to care for the environment. It changed me. I had, as John Wesley would say, a "warming of my heart"… a conversion to a cause which I believe every Christian should be committed to. And so, yes, I had a change that only God could do. Only God could affect that kind of change in my heart. Because I am a stubborn person.

Q: Was it an advantage to change your ways of living?

Rev. Cizik:
Absolutely. I came back from the Oxford conference on climate change and talked to my wife, Virginia. I said, *"Ginny, we're gonna have to change some things."* The first thing we did was sell our recreational vehicle. I decided it just wasn't a good use of resources. And it was a gas guzzler, besides. So we bought a Prius, a hybrid. Because it was not only a gas saver, but it was economically and environmentally the right thing to do. And we've taken a look at recycling. And then we've done something that I would never have imagined – we've begun to enjoy the environment that God created, in new ways, with our children. Recognizing that this is what God wants us to do. And in so doing we've got a different kind of appreciation for the world that God has created

Q: What should pastors be saying to their congregation in terms of a sermon?

Rev. Cizik:
I think pastors should begin by teaching the word of God. This is God's world. When we destroy or deplete it, we violate his will. I think that is the beginning. But there are also concrete actions which our believers in churches can take, from recycling to looking at the energy that our churches use, to looking for energy-efficient ways of operating our congregations.

Q: Can evangelical Christians and environmentalists work together to effect change?

Rev. Cizik:
Absolutely. If we can work with Tibetan Buddhists to pass effective international religious freedom legislation, then surely we can work with our own fellow environmentalist, many of whom are Christians, to save the Earth. After all, God calls us to do this. So, if one man's reason for doing it is good science and another man's reason for doing is good faith, then so be it and all the better.

Q: What concerns you the most?

Rev. Cizik:
I honestly fear that I would fail to be obedient to the command that God has given me and my role here in Washington. That given the authority I have, entrusted by 54 denominational leaders and church pastors from around the country, that I fail to heed God's call to effectively influence our nation's public policies. That is the mandate I have been given and entrusted with. That is the greatest fear I have. Because God doesn't intend to ask me *"Rich, how did I create the Earth?"* He won't ask me that. He'll say *"Rich, what did you do to protect that which I created?"* And that is an awesome question that deserves a good answer.

Q: That is a wonderful thought.

Papal Messages on Climate Change

Statements by His Holiness Pope Francis I

Pope Francis I on "Catholics as Custodians of Creation"

Pope Francis today made the religious case for tackling climate change, calling on his fellow Christians to become "Custodians of Creation" and issuing a dire warning about the potentially catastrophic effects of global climate change.

His Holiness said that respect for the "beauty of nature and the grandeur of the cosmos" is a Christian value, noting that failure to care for the planet risks apocalyptic consequences.

"Safeguard Creation," he said. "Because if we destroy Creation, Creation will destroy us! Never forget this!"

Vatican statement, May 21, 2014

Pope Francis I's Message to President Correa of Ecuador

"Take good care of creation. St. Francis wanted that. People occasionally forgive, but nature never does. If we don't take care of the environment, there's no way of getting around it."

Vatican City, April 22, 2013

Pope Francis I: Statement to the Diplomatic Corps

I wish to mention another threat to peace, which arises from the greedy exploitation of environmental resources. Even if "nature is at our disposition," all too often we do not "respect it or consider it a gracious gift which we must care for and set at the service of our brothers and sisters, including future generations." Here too what is crucial is responsibility on the part of all in pursuing, in a spirit of fraternity, policies respectful of this earth which is our common home.

I recall a popular saying: "God always forgives, we sometimes forgive, but when nature – creation – is mistreated, she never forgives!"

Vatican City, January 14, 2014

Pope Francis I: Homily in Saint Peter's Square

The vocation of being a "protector," means protecting all creation, the beauty of the created world, as the Book of Genesis tells us and as Saint Francis of Assisi showed us.... In the end, everything has been entrusted to our protection, and all of us are responsible for it. Be protectors of God's gifts!

Whenever human beings fail to live up to this responsibility, whenever we fail to care for creation and for our brothers and sisters, the way is opened to destruction and hearts are hardened.

Please, let us be "protectors" of creation, protectors of God's plan inscribed in nature, protectors of one another and of the environment. But to be "protectors", we also have to keep watch over ourselves! Let us not forget that hatred, envy and pride defile our lives! Let us protect with love all that God has given us!

Vatican City, March 19, 2013

Papal Messages on Climate Change

Statements by His Holiness Pope Benedict XVI

Pope Benedict XVI Addresses Global Climate Change

Pope Benedict XVI appealed for the success of a UN climate change conference that is opening tomorrow in Durban, South Africa. Speaking to the faithful gathered in St Peter's Square for the Sunday Angelus prayer, Pope Benedict expressed the hope that "all members of the international community might reach agreement on a responsible, credible response," to the phenomenon of climate change, which he described as "complex and disturbing."

The Holy Father also asked that leaders' response be consonant with the spirit and requirements of solidarity, taking into account the needs of the poorest people and future generations. The Holy Father's appeal came after the traditional prayer of Marian devotion, ahead of which he spoke about the new liturgical season of Advent, which began this weekend with First Sunday Vespers. It was a theme to which he returned in his English-language remarks, as well.

Letter, November 27, 2011

Pope Urges International Agreement on Climate Change

Pope Benedict XVI addressed delegates of 194 countries gathering in Durban, South Africa for the latest round of international climate change negotiations. He urged a strong global agreement to address the challenge of climate change: "I hope that all members of the international community can agree on a responsible, credible and supportive response to this worrisome and complex phenomenon, keeping in mind the needs of the poorest populations and of future generations."

Letter, November 25, 2011

Pope Benedict XVI: "The Listening Heart: The Foundations of Law"

"I would say that the emergence of the ecological movement in German politics since the 1970s, while it has not exactly flung open the windows, nevertheless was and continues to be a cry for fresh air which must not be ignored or pushed aside, just because too much of it is seen to be irrational. Young people had come to realize that something is wrong in our relationship with nature, that matter is not just material for us to shape at will, but that the earth has a dignity of its own.... In saying this, I am not promoting any particular political party – nothing could be further from my mind. If something is wrong in our relationship with reality, then we must all reflect seriously on the whole situation and we are all prompted to question the very foundations of our culture. Allow me to dwell a little longer on this point. The importance of ecology is no longer disputed. We must listen to the language of nature and we must answer accordingly.

The Reichstag, Berlin, Germany, Sept 22, 2011

Vatican Releases Major Climate Change Report

A working group of the Pontifical Academy of Sciences, one of the oldest scientific institutes in the world, issued a sobering report (May 6, 2011) on the impacts for humankind as a result of the global retreat of mountain glaciers as a result of human activity leading to climate change.

In their declaration, the working group calls, "on all people and nations to recognize the serious and potentially irreversible impacts of global warming caused by the anthropogenic emissions of greenhouse gases and other pollutants, and by changes in forests, wetlands, grasslands, and other land uses."

May 6, 2011

Pope Benedict calls attention to "the problems associated with climate change"

In a wide-ranging, prophetic, and challenging World Day of Peace message on climate change and environmental justice, Pope Benedict recalled our biblical tradition, highlighted teachings from previous popes and implored us to rethink the path which we are traveling together.

"Can we remain indifferent before the problems associated with such realities as climate change, desertification, the deterioration and loss of productivity in vast agricultural areas, the pollution of rivers and aquifers, the loss of biodiversity… and the growing phenomenon of 'environmental refugees?' If You Want to Cultivate Peace, Protect Creation."

2010 World Day of Peace Message

Pope Benedict XVI Sends Powerful Message to Climate Summit

Pope Benedict XVI sent a compelling message to those gathering at the 2009 Summit on Climate Change when he conveyed his "support to leaders of governments and international agencies who will meet at the United Nations to discuss the urgent issue of climate change…. "How important it is then, that the international community and individual governments send the right signals to their citizens and succeed in countering harmful ways of treating the environment!… The protection of the environment, and the safeguarding of resources and of the climate, oblige all leaders to act jointly, respecting the law and promoting solidarity with the weakest regions of the world."

Vatican City, 2009

Pope Benedict XVI: Environmental Problems are Reaching Crisis Proportions

The gradual depletion of the ozone layer and the related 'greenhouse effect' has now reached crisis proportions as a consequence of industrial growth, massive urban concentrations and vastly increased energy needs. Industrial waste, the burning of fossil fuels, unrestricted deforestation, the use of certain types of herbicides, coolants and propellants: all of these are known to harm the atmosphere and environment. The resulting meteorological and atmospheric changes range from damage to health to the possible future submersion of low-lying lands.

World Youth Day, June 2008

Pope Benedict XVI Urges Agreement at COP-15, Copenhagen

"I share the growing concern caused by economic and political resistance to combating the degradation of the environment.... If we wish to build true peace, how can we separate, or even set at odds, the protection of the environment and the protection of human life, including the life of the unborn? It is in man's respect for himself that his sense of responsibility for creation is shown."

Vatican City, 2009

Holy Father's Easter Message Highlights Climate Change

Pope Benedict XVI links the threat of climate change with so many other threats to life and dignity in his annual Easter message:

"At a time of world food shortage, of financial turmoil, of old and new forms of poverty, of disturbing climate change, of violence and deprivation which force many to leave their homelands in search of a less precarious form of existence, of the ever-present threat of terrorism, of growing fears over the future, it is urgent to rediscover grounds for hope. Let no one draw back from this peaceful battle that has been launched by Christ's Resurrection. For as I said earlier, Christ is looking for men and women who will help him to affirm his victory using his own weapons: the weapons of justice and truth, mercy, forgiveness and love."

"The brutal consumption of Creation begins where God is not, where matter is henceforth only material for us, where we ourselves are the ultimate demand, where the whole is merely our property and we consume it for ourselves alone...I think, therefore, that true and effective initiatives to prevent the waste and destruction of Creation can be implemented and developed, understood and lived, only where Creation is considered as beginning with God."

Urbi et Orbi, Easter 2009

* * *

From Roman Catholic Cardinal Wilfred Napier, Durban, South Africa

Failure to Act on Climate Change will Cause Terrifying Human Suffering

We call upon our leaders, those of our faiths, and all people of Earth to accept the reality of the common danger we face from global climate change, [and] the imperative and responsibility for immediate and decisive action and the opportunity to change.

We express our displeasure with international political leadership which has failed to take decisive steps to make the changes required for the survival of humanity and life on earth.

We demand that our political leaders honor previous commitments and move towards ethically responsible positions and policies.... Without deep emissions cuts now, dangerous global warming will occur. It will cause human suffering on a terrifying scale.

Christianity Today, December 3, 2011

Roman Catholic Initiatives on Global Climate Change

by Fr. Jacek Orzechowski, OFM
May 20, 2013

My name is Fr. Jacek Orzechowski and I'm a Franciscan friar and Catholic priest.

In 2001, Pope John Paul II, warned that humanity was moving toward a catastrophe. He said: "Man is no longer 'minister' of the Creator, but rather an autonomous despot who is understanding that he must finally stop before the abyss." That same year, the US Catholic Bishops issued a statement entitled: "**Global Climate Change: A Plea for Dialogue, Prudence and the Common Good.**"

For the bishops, global climate change is, first and foremost, a religious and moral issue. In their statement, they offer several themes and values rooted in the Catholic social teachings.

The first theme is the universal common good. The earth's atmosphere belongs to God. We must share it with all human and non-human creatures and with the future generations.

The second theme is Stewardship of God's Creation. Rejecting consumerism, addressing the systemic injustice in our society, fulfilling one's civic responsibilities – these are some of the essential elements of what it means to be a faithful steward of God's creation.

The third theme highlighted by the bishops is the concern for future generations. We ought to consider the impact of our action on those who have not yet been born. What we do – or fail to do – often have profound moral implications.

The forth theme is authentic development. Human beings are far more than what they can produce and consume. The bishops call for a re-examination of the dominant model of progress in our society that fuels greed and leaves little space for values such as solidarity and altruism. In a similar vain, the bishops warn about the dangers of voracious consumerism. They insist that our society must recognize the limits and hazards of growth.

Furthermore, to be able to effectively deal with the problem of climate change – the bishops assert - we must seek to reduce poverty and improve education and social conditions for women.

The fifth and the last theme is: caring for the poor and issue of equity. The Unites States and other wealthier nations have a moral responsibility to help impoverished countries to adapt to and mitigate the effects of climate change.

The Catholic bishops' statement on climate change led to a creation of Catholic Coalition on Climate Change (Franciscan Action Network is one of the members of this coalition).

This initiative has been a very important step in engaging the Catholic parishes, schools and individual people in learning about the religious and moral dimension of climate change and spurring them into action. Catholic Coalition on Climate Change is making a difference, and these are some of the examples:

- Thousands of people of faith and 16 Catholic Universities signed the Saint Francis Pledge to Care for Creation and the Poor. This includes a commitment to reduce one's individual and collective carbon footprint.

- Last summer over 25,000 people watched a documentary *Sun Come Up* about the plight of some of the world's first "climate refugees," and participated in the education efforts.

Furthermore, this fall, Catholic Coalition on Climate Change will be launching its new educational campaigns called: "**The Climate Crisis is a Crisis of the Sabbath**" and "**Melting Ice, Mending Creation.**"

Next year, the Coalition will be focusing on creating a green loan funds and on the efforts to significantly reduce a collective carbon footprint of Catholic institutions.

The Catholic bishops' statement on climate change has also helped to incentivize organizations such as Franciscan Action Network to become a stronger advocate for alternative energy such as wind power and to vigorously oppose misguided and immoral projects such as Keystone XL pipeline that perpetuate our society's deadly addiction to dirty fossil fuels. Franciscan Action Network has just launched Franciscan Earth Corps - a national effort to engage young adults in the care for God's creation.

Pope Benedict XVI said that we should not to remain on the sidelines in the struggle or justice. That includes environmental justice. During his pontificate, the Vatican became the first nation in the world to become carbon neutral. Benedict XVI also warned that the earth was speaking to us and we must listen and decipher her message, if we wanted to survive.

Pope Benedict's successor, has taken the name of Francis – after St. Francis of Assisi – the Patron Saint of ecologists. Pope Francis continues to call all people of good will to safeguard and care for creation and urges them not to succumb to "the worship of the golden calf; that is, the cult of money and the dictatorship of an economy which is faceless and lacking any truly human goal."

Given the gravity and urgency of global climate change, let me echo the words from St. Francis of Assisi who at the end of his life and ministry of compassion toward the poor said in all humility:

"Brothers and sisters, let us begin anew,
for up until now, we have done nothing."

Protestant Reflections on Climate Change and Caring for Creation

by Reverend Olin M. Ivey, Ph.D
Tennessee Earth Care
Chattanooga, Tennessee
(abridged)

MOST DENOMINATIONS HAVE BEAUTIFULLY CRAFTED SOCIAL ACTION STATEMENTS on climate change, its sources, and the Church's responsibility to address its causes and its solutions.

The human's sinful condition is equally personal and social. Sinfulness requires repentance. Repentance leads to a concerted effort toward a new understanding of life based in continual inquiry and reflection and leading to new social structures and corrective, far-sighted actions that combine justice and love. The scientific method and the findings of science inform our thoughts and actions. When 97% of the world's scientists say human activity, led by the use of fossil fuels, constitutes the main cause of rising temperatures and many sound alarm bells, global warming increasingly becomes a concern of mainline Protestant churches.

Covenant forms the meeting place of humanity's faith, God's love, and the Earth and all its creatures. The human's loving responsibility extends to the most far-flung reaches of the universe and to the minutest of creatures yet undiscovered. All creatures play a vital role in the web-of-life. An adherence to the Commons and the common good is vital to the continuing vibrancy of this covenantal troika.

Jesus teaches us that when greed reigns, it engulfs one's being and controls one's actions. It breaks covenant. The same is true for institutions and corporations. In this case, it is not "moral man and immoral society" but immoral individuals; immoral society.

Global warming's main source of rising CO2 emits from our ever-growing worldwide addiction to fossil fuels. We have now passed the milestone atmospheric level of 400 ppm CO_2 concentration as recorded this May at the Mauna Loa Observatory in Hawaii, and the level continues to rise. This is its highest level in *three million years.* We call for governments and corporations around the world... to admit - just *admit* - the devastating affects of heat-trapping carbon dioxide in the atmosphere. Furthermore, repent, quit the "fossil fuel habit," and begin immediately the transition into renewable energy and other energy efficient devices:

- No longer willfully ignore scientific data
- No longer intentionally foster destructive activities on the Earth's land, water, and air
- No longer wage deceptive, multi-layered campaigns to deceive the public
- No longer entice politicians to do your destructive bidding

Quit it! In God's name, stop it! Or by any other rationale or emotion, cease!

As disciples of Jesus, outrage toward the irresponsibility of others must be matched by our own searing introspection. Protestants are far better at proclamation than at eco-friendly action. But as disciples of Jesus, we are to develop sustainable practices, modeling a responsible, ecological approach in areas of our life together, internally as a congregation and externally as co-inhabitants with other creatures - including those with whom we are outraged. Enmeshed in the dirt of the Earth, we are thrown together as participants with all others, but in discipleship, we are intentional stewards by developing a comprehensive vision. How does that happen? By transforming our minds through education, our spirits through worship, our hearts and hands through sustainable practices, and our resources through targeted planning. It is a constant journey of "consciousness-raising" and "conscientiousness-quickening."

This is God's good, green Earth. We must not destroy it - as we certainly will if we continue making blinder, oil-centric decisions - and the destruction of human civilization will come sooner than most think.

God, the source of transformation, placed freedom at the heart of the universe - freedom to choose which way to go. The lure of God is always toward the good. Our responsibility is to follow the common good, the daily goodness discovered in an horizontal loving and caring for all of life.

The Protestant church sees the Earth as a "faith garden." In faith we care for the Earth as a garden in symbiotic relationship with all the other creatures. In concert with God, we seek to sustain the Earth by learning to think and act like nature. We learn from nature the intertwining, supportive relationship that God has established between all the elements of the Earth. Our activities transform the Earth into support systems of our human culture but they must be done in a respectful manner toward nature - the soil, the air, the water, the minerals and all the creatures everywhere.

How do we do this? Here are ten meager steps on the journey of creation care:

First, as people of faith, we continue to be grounded in scripture and temper that by tradition, experience, reason, wonder, reverence, and nature itself.

Second, We celebrate the beauty and resourcefulness of the Earth. So much of it is like a Garden of Eden in its abundance of species and magnificence. Within creation God shows us the way to live sustainably. We dedicate ourselves to loving, faithful action on behalf of all of creation.

Third, using a number of categories, we audit and record where we are in terms of a green index within the congregation and in our daily lives. Data, attitudes, and behaviors – tracking them in each of these spheres – are required to measure the paradigm shift that has to take place. We formulate a management program to guide us. We benchmark our progress. We become an "Earth Church" and extend that into our everydayness. We reduce our footprint upon the Earth. We understand that any action that affects the systems and species of the earth is right only if it conserves, preserves, or restores the stability, integrity, viability, beauty, and fruitfulness of what God has created.

Fourth, we seek out others who are on this same journey. We look for resources from other churches, from governmental agencies, and from other groups and individuals. Through God's graciousness, we invite others to transform their lives and their communities and to join us on our journey of restoring the Earth and its systems.

Fifth, we play in the garden: a pot, a 4 x 4 designated area, a community garden, a backyard garden, a garden at church. This provides an opportunity to observe, to hear, and to converse with creation up close. A gardener deals with many other species. Those species are dealing with their own lives, trying to eek out a living. Cross fertilization occurs. New life happens – sometimes when it is not a part of the plan. Unexpected support systems and cooperation mingle with the struggle to survive. A rhythm and a rhyme emerge in this dialogue with nature and within nature.

Sixth, we cook. Life in all its fullness is discovered in understanding our foods, good diets, the very activity of preparing the ingredients and blending them together, and then the celebration of life in the breaking of bread together. Some Protestant denominations have made food central to the faith and life journey. When that is central, care for the earth is also central.

Seventh, we seek justice and "do justice" by reversing the devastating effects that global warming has had upon the Earth and its people, especially the poor and those whose lives are riddled with poverty *and* prejudice. In concert with Jesus's concern for "the least of these," we seek sensitivity to the plight of vulnerable peoples and for threatened and endangered species. Reverence for all of humanity as well as the whole of creation blossoms into an urgency for justice. We reverse our habits of unsustainable energy and consumption. Sufficiency in living is sufficient for life. We trace the unjust consequences of own lifestyles and the mindless choices that threaten other people and natural systems. We strive to reverse this by straightening out our priorities. We ask for God's help in our continual journey of transformation and commitment. We realize that this is no "lifestyle adjustment" but a radical paradigmatic shift in who we are as spiritual beings. Our actions must always preserve, protect, and restore the panoply of nature: clean water, clean air, the oceans, the forests, and creature habitat. Thankful for our redemption, we seek healing for others and the Earth.

Eighth, we walk humbly with God, tread lightly upon creation in a posture of humility, and seek to shed ourselves of arrogance and greed through being kind, merciful, and loving.

Ninth, we take political and economic action. We tell the truth about God's creation and the moral, ethical imperative that drives us to do what we can to be God's servants for the Earth and to speak that Truth to power.

But that's where we started this discussion. Yet, there is one more:

Tenth, as resurrection people empowered by the Holy Spirit, our lives are enthused with hope. We help bring the Earth back from the brink of severe, human-induced warming and toward ecological sustainability. We tell the truth about God's creation. No matter how many set backs, we know that this planet is God's good Earth. It is that which sustains.

Not every church is doing these things. By far, more are not than are involved at this juncture. But it's happening. Some Protestant churches can be found doing at least a few of these. Every now and then, some churches really get what it means to be an "Earth Church" filled with Earth Pilgrims who are on an Earth Journey, and who call to the world to join them.

Shalom happens! Praise be to God!

Reverend Olin Ivey is a Methodist minister in Chattanooga, Tennessee.

An Open Letter to COP-17

by His All-Holiness Ecumenical Patriarch Bartholomew

Durban, South Africa
November 28, 2011

IT IS WITH SINCERE DELIGHT THAT WE GREET THE distinguished governmental leaders, eminent dignitaries and honorable participants of the COP-I7 meeting in Durban, South Africa. You have assembled at a critical time for a vital opportunity to address the serious challenges of the environmental crisis.

Throughout the conference, you will be presented with scientific, economic, and social considerations of the problems that we face. You will hear of regional and global issues related to troubling changes in our climate, including extreme weather conditions, depletion of food supplies and agriculture, as well as escalating diseases on our planet.

The manifold effects are widely known and well substantiated. The various statistics are abundantly alarming and easily available. Nevertheless, these are, at least in our mind, just the beginnings of the disruption to the climate system. Moreover, and more importantly, they are merely the surface of the problem.

> *Global climate change presents an unprecedented threat to... life on earth.... We have already denounced ecological abuse as sin against God; we should recognize how it is also a crime against humanity.*

As we look behind and beyond these first effects, unfortunately we tend to ignore... the lives of those of our brothers and sisters who are and will be directly affected by these conditions and consequences. We conveniently tend to overlook... that we are all responsible for the future of our planet and for human life. Climate change affects all people and all nations. None among us can remain a mere spectator.

Previous meetings in Copenhagen and Cancun failed to produce a legally binding agreement with regard to affluent and poor nations. Furthermore, next year marks the expiration of the Kyoto Protocol, even if prominent and prosperous nations have yet to submit to measurable mitigation and adaptation plans.

Beloved friends, at first glance, it may appear strange that a religious institution concerned with "sacred" values can be so profoundly involved in "worldly" issues. Yet, there is much more to climate change than environmental preservation. We are dealing with a profoundly moral and spiritual problem. Our ministry and mandate is to sensitize consciences and energize listeners of good will. Therefore, we ask that you contemplate the following challenges:

(i) A first challenge for participants at this conference is *the struggle to surmount national and regional considerations and to consider the larger picture.* Climate change is a global problem. We share one world and the same resources, one atmosphere and the same habitat. We are inseparably interconnected. Any genuine solution demands the ability to think for the whole world. We are all connected and our actions affect each other. Conservation and compassion are intimately interrelated.

(ii) A second challenge is *remembering that sacrifice is needed to arrive at a successful conclusion.* When will we face the inevitable truth that all ecological activity is ultimately judged by its impact on the poor? When will we sense the painful reality that the continent that has scarcely contributed to global warming is bearing the most detrimental repercussions, even while being the least equipped to cope with its consequences? The greatest delusion is that measures to deal with climate change must not... affect economic growth. Without sacrifice... we cannot reach the unity necessary for an enduring agreement.

(iii) A third challenge is *the priority of securing moral leadership.* Global climate change presents an unprecedented threat to the integrity and diversity of life on earth. At the Ecumenical Patriarchate, we have already denounced ecological abuse as sin against God; we should recognize how it is also a crime against humanity. Blame is no solution. Instead, we must discover the resources that lie deep within the human spirit in order to develop a sense of urgency and resolve.

The moral leadership that is required is a commitment to embrace and become the solutions that we advocate. We humbly invite all of you - whether delegates, politicians, activists, and individual citizens - to make a personal commitment to effect transformations in the many and minute details of daily life, especially in the way we deal with energy and relate to the poor.

The world is watching. And the world is waiting. May God bless your deliberations.

At the Ecumenical Patriarchate, the 23rd of November 2011
Prayerfully yours,

BARTHOLOMEW
Archbishop of Constantinople-New Rome, and Ecumenical Patriarch

Climate Change: A Rabbi Speaks Out

by Rabbi Warren G. Stone
Temple Emanu El
Kensington, Maryland

AN ANCIENT JEWISH MIDRASH TEACHES that when God took Adam around the Garden of Eden and showed him its magnificence and splendor, God spoke to him saying, "If you destroy it, there is no one else besides you [to restore it]!"

Those words ring mightily today, for the very future of life as we know it is at stake. I fervently believe that climate change, with the destruction that it is wreaking on our fragile, sacred earth, has become the most profound religious issue of our times. Like Adam, we have been warned and cannot plead ignorance; like Adam, will we fail to heed God's words?

Who is responsible for responding to the challenge of global climate change? We tend to think that it is the scientist, the statesman and the environmentalist upon whom this responsibility lies. But climate change is an urgent moral and spiritual issue for all people.... We are witnessing its impact right now, and we can foresee the havoc it will wreak on the health and survival of further generations. The future will bring environmental refugees in numbers unknown in previous ages. As a result of climate change and habitat destruction, a myriad of species now faces a silent genocide.

> *I fervently believe that climate change, with the destruction that it is wreaking on our fragile, sacred earth, has become the most profound religious issue of our times.*

As a Rabbi and religious leader, I am concerned about our common future, the quality of life for our families and the threatened species of our world, including our own. I join fellow religious leaders in that concern. But it is not enough to care about climate change, forest devastation and environmental threats to clean water, air and seas. It is incumbent upon every religious leader, religious institution and person of faith to serve as beacons to our communities, illustrating by our actions and example our spiritual commitment to our earth and its threatened and limited resources.

In a world where matters of faith seem so often and so tragically to divide us, there is no issue that aligns us more deeply than our shared dependence upon and sacred responsibility to this tiny planet, enfolded within its fragile atmosphere, spinning in the vastness of time and space. I experienced this shared conviction most profoundly, when in 1997, I served as the Jewish NGO representative at the United Nations climate talks in Kyoto, Japan. I met with Catholic, Protestant, Muslim, Hindu and Buddhist leaders from around the world. We spoke at Kyoto's largest Buddhist Temple, and all concurred that our human actions, our sins, have damaged the environment. Each speaking from the voice of his or her own authentic spiritual tradition, we affirmed our religious responsibility to act. Amidst Buddhist chanting, I blew the shofar, a ram's horn, the blast of sound that has been Judaism's ancient call to action since the days we wandered, searching for our way, in the desert.

I carried this profound experience back to my own country and my own community. Here, too, I found that faith traditions can readily unite on issues of climate change. Working for many years with the National Partnership on Religion and the Environment, I have joined interfaith leaders to lobby on Capitol Hill and to meet with White House staff. Political leaders are eager to hear our religious point of view. Statements by Catholic Bishops, Protestants leaders, Rabbis and Tribal Leaders have symbolic power and carry political weight. Formal resolutions affirmed by hundreds of thousands of persons of faith help embolden our legislators to act. This year, religious leaders stood with sympathetic legislators on the U.S. Capitol's steps, raising our voices to stop the drilling in the Arctic Wildlife refuge. The opportunity to be heard is greater than in previous decades, and we have a prophetic responsibility to seize it. Bold initiatives are needed – and needed now – to protect species, to create incentives for the development of alternative energies, to protect endangered coastal areas and to mitigate our dependence on fossil fuels.

> *Bold initiatives are needed – and needed now – to protect species, to create incentives for the development of alternative energies, to protect endangered coastal areas and to mitigate our dependence on fossil fuels.*

Of course, our collective, interfaith efforts gather their strength from the work each of us does within our own particular communities. As chair of the Environmental Committee of the Central Conference of American Rabbis, I have joined with many committed colleagues to use our faith tradition to increase awareness and encourage action in response to climate change and other environmental challenges. We have passed national resolutions on climate change and energy policy and have established environmentally conscious guidelines for our myriad congregations around the country. For example, we recently celebrated Chanukah, the Jewish holiday of light, renewal and commemoration of bold action that honors one's faith; during the holiday, we mounted a very successful COEJL (Coalition on the Environment and Jewish Life) national campaign – "Let There Be (Renewable) Light! or How Many Jews Does it Take to Change a Light Bulb?" Thousands of congregations encouraged their members to install energy saving, compact flourescent light (CPF) bulbs in their synagogues and homes and to add to their holiday ritual a ceremony that calls this generation to environmental action, in response to the moral imperatives of our own times.

And finally, I believe that our religious voice must be strongest closest to home, manifest in how we daily live The congregation I serve, Temple Emanuel of the Greater Washington area, has worked on greening its agenda for 18 years. We believe that local action by religious communities can have a national and international impact.

The congregation I serve has installed solar panels on the roof for our eternal light, added wind power from a regional collective, made use of energy efficient zoning, lighting and office equipment and in our building made use of passive solar throughout the building. We planted a sustainable garden to meet our annual ritual needs, growing grapes, horseradish, and indoor olive and pomegranate trees. We regularly schedule environmental Shabbats and other opportunities for learning with our state representative and national leaders. We sell CPF bulbs and have information about climate change on our coffee tables. We have become an EPA Energy Star community and one of the nation's first "zero carbon footprint" communities by supporting Carbonfund.org and their alternative energy investments.

Our web page – www.templeemanuelmd.org – includes our Green Shalom action guide which is designed to educate and spur further community involvement and environmental action in our own homes and community. This community focus has borne fruit, with a good number of our young people choosing science, media, religion and public policy arenas that deal directly with environmental issues.

I believe that our religious voice must be strongest closest to home, manifest in how we daily live.

There is so much that each of us can and must do, within our own homes, congregations and communities, and beyond, as we work together, in common cause, to preserve and sanctity life. Religious communities have a crucial moral role in affirming the profound need to engage on the issue of climate change.

As RabbiTarphon of the second century reminds us: "It is not your duty to finish all the work, but neither are you are liberty to desist from it." May it be that years hence, our children and our children's children will look back with appreciation to this moment when we heeded one of the great moral imperatives of our time.

May they know that we had the vision and the strength to fulfill our sacred obligation to preserve and protect the earth in all of its majesty, this garden with which we have been entrusted, for those who will follow.

Why Should Evangelical Christians Care About Climate Change?

Here are five reasons from an Evangelical Christian climate scientist

An Interview with Climate Scientist Dr. Katharine Hayhoe
May 13, 2014

by Chris Mooney
(This interview originally appeared in *Mother Jones magazine*)

CLIMATE SCIENTIST KATHARINE HAYHOE, AN EVANGELICAL CHRISTIAN, has had quite a run lately. She was recently featured in the first episode of the TV series *The Years of Living Dangerously*, meeting with actor Don Cheadle in her home state of Texas to explain why faith and a warming planet aren't in conflict. Then *Time* magazine named her one of the 100 most influential people of 2014.

Why is Hayhoe in the spotlight? Simply put, millions of Americans are evangelical Christians, and their belief in the science of global warming is well below the national average. If anyone has a chance of reaching this vast and important audience, Hayhoe does. "I feel like the conservative community, the evangelical community, and many other Christian communities, I feel like we have been lied to," explains Hayhoe. "We have been given information about climate change that is not true. We have been told that it is incompatible with our values, whereas in fact it's entirely compatible with conservative and with Christian values."

Hayhoe's approach to science—and to religion—was heavily influenced by her father, a former Toronto science educator and also, at one time, a missionary. "For him, there was never any conflict between the idea that there is a God, and the idea that science explains the world around us," says Hayhoe. When she was 9, her family moved to Colombia, where her parents worked as missionaries and educators, and where Hayhoe saw what environmental vulnerability really looks like. "Some of my friends lived in houses that were made out of cardboard Tide boxes, or corrugated metal," she says. "And realizing that you don't really need that much to be happy, but at the same time, you're very vulnerable to the environment around you, the less that you have."

> *The choices we make today and over the next decade will have a radical impact on the path we travel in the future.*

"In terms of addressing the climate issue," says Hayhoe, "we don't have time for everybody to get on the same page regarding the age of the universe."

Her research today, on the impacts of climate change, flows from those early experiences. And of course, it is inspired by her faith, which for Hayhoe puts a strong emphasis on caring for the weakest and most vulnerable among us. "That gives us even more reason to care about climate change," says Hayhoe, "because it is affecting people, and is disproportionately affecting the poor, and the vulnerable, and those who cannot care for themselves."

The fact remains that most evangelical Christians in the U.S. do *not* think as Hayhoe does. Recent data from the Yale Project on Climate Change Communication suggests that while 64 percent of Americans think global warming is real and caused by human beings, only 44 percent of evangelicals do. Evangelicals in general, explains Hayhoe, tend to be more politically conservative, and can be quite distrusting of scientists (believing, incorrectly, that they're all a bunch of atheists). Plus, some evangelicals really do go in for that whole "the world is ending" thing—not an outlook likely to inspire much care for the environment. So how does Hayhoe reach them?

From our interview, here are five of Hayhoe's top arguments, for evangelical Christians, on climate change:

1. Conservation is conservative

The evangelical community isn't just a religious community, it's also a politically conservative one on average. So Hayhoe speaks directly to that value system. "What's more conservative than conserving our natural resources, making sure we have enough for the future, and not wasting them like we are today?" she asks. "That's a very conservative value."

Indeed, many conservatives don't buy into climate science because they don't like the "big government" solutions they suspect the problem entails. But Hayhoe has an answer ready for that one too: Conservative-friendly, market-driven solutions to climate problems are actually all around us. "A couple of weeks ago, Texas… smashed the record for the most wind energy ever produced. It was 38 percent of our energy that week, came from wind," she says. And Hayhoe thinks that's just the beginning: "If you look at the map of where the greatest potential is for wind energy, it's right up the red states. And I think that is going to make a big difference in the future."

2. Yes, God *would* let this happen

One conservative Christian argument is that God just wouldn't let human activities ruin the creation. Or, as Sen. James Inhofe of Oklahoma has put it, "God's still up there, and the arrogance of people to think that we, human beings, would be able to change what he is doing in the climate, is to me, outrageous."

Hayhoe thinks the answer to Inhofe's objection is simple: From a Christian perspective, we have free will to make decisions and must live with their consequences. This is, after all, a classic Christian solution to the theological problem of evil. "Are bad things happening? Yes, all the time," says Hayhoe. "Someone gets drunk, they get behind the wheel of a car, they kill an innocent bystander, possibly even a child or a mother."

Climate change is, to Hayhoe, just another wrong, another problem, brought on by flawed humans exercising their wills in a way that is less than fully advisable. "That's really what climate change is," she says. "It's a casualty of the decisions that we have made.

3. The Bible does not approve of letting the world burn

Hayhoe agrees with the common liberal perception that the evangelical community contains a significant proportion of apocalyptic or end-times believers—and that this belief, literally that judgment is upon us, undermines their concern about preserving the planet. But she thinks there's something very wrong with that outlook, and indeed, that the Bible itself refutes it.

"The message that we don't care about anybody else, and let the world burn, that message is not a consistent message in the Bible," says Hayhoe. In particular, she thinks the apostle Paul has a pretty good answer to end-times believers in his second epistle to the Thessalonians. Hayhoe breaks Paul's message down like this: "I've heard that you've been quitting your jobs, you have been laying around and doing nothing, because you think that Christ is returning and the world is ending." But Paul serves up a rebuke. In Hayhoe's words: "Get a job, support yourself and your family, care for others—the poor and the vulnerable who can't care for themselves—and do what you can to make the world a better place, because nobody knows when that's going to happen."

4. Even if you believe in a young Earth, it's still warming

One reason there's such a tension between the evangelical community and science is, well, science. Many evangelicals are young-Earth creationists, who believe that the Earth is 6,000 or so years old.

Hayhoe isn't one of those. She studied astrophysics and quasars that are quite ancient; and as she notes, believing the Earth and universe to be young creates a problematic understanding of God: "Either you have to believe that God created everything looking as if it were billions of years old, or you have to believe it is billions of years old."

But when it comes to talking to evangelical audiences about climate change, Hayhoe doesn't emphasize the age of the Earth, simply because, she says, there's no need. "When I talk to Christian audiences, I only show ice core data and other proxy data going back 6,000 years," says Hayhoe, "because I believe that you can make an even stronger case, for the massive way in which humans have interfered with the natural system, by only looking at a shorter period of time."

"In terms of addressing the climate issue," says Hayhoe, "we don't have time for everybody to get on the same page regarding the age of the universe."

5. "Caring for our environment is caring for people"

Finally, Hayhoe thinks it is crucial to emphasize to evangelicals that saving the planet is about saving people ... not just saving animals.

"I think there's this perception," says Hayhoe, "that if an environmentalist were driving down the road ... and they saw a baby seal on one side and they saw a human on the other side, they would veer out of the way to avoid the baby seal and run down the human." That's why it's so important to emphasize how climate change affects *people* (a logic once again affirming the perception that the polar bear was a terrible symbol for global warming). And there's bountiful evidence of this: The just-released Intergovernmental Panel on Climate Change's "Working Group II" report on climate impacts emphasizes threats to our food supply, a risk of worsening violence in a warming world, and the potential displacement of vulnerable populations.

So is the message working? Hayhoe thinks so. After all, while only 44 percent of evangelicals may accept modern climate science today, she notes that that's considerable progress from a 2008 Pew poll, which had that number at 34 percent.

Ultimately, for Hayhoe, it comes down to this: "If you believe that God created the world, and basically gave it to humans as this incredible gift to live on, then why would you treat it like garbage? Treating the world like garbage says a lot about how you think about the person who you believe created the Earth."

An Open Letter to Clergy

The Reverend Thomas Carr
First Baptist Church, West Hartford, Connecticut
November 6, 2012

 As I PONDER THE AFTERMATH OF THE HURRICANE that left massive devastation on the East Coast, I am saddened and grieved over such loss. My heart breaks for those who suffered and are still struggling in the aftermath. I, like you, am seeking to do whatever I can to assist those in their time of immediate need. So many people have suffered so much and it will take months, if not years, to rebuild.

At the same time, I am angry and perplexed. For two decades we have known the reason why storms like these are getting more and more intense, why wildfires are increasing, why droughts are becoming more permanent and floods are increasing in duration and intensity: climate change and the primary reason the climate is changing radically and rapidly is our way of life built on the use and burning of fossil fuels. I'm angry because there has been an intentional, highly organized and more than well financed campaign to confuse Americans and thus delay making the changes needed to begin addressing this critical problem.

I'm perplexed as to why we preachers have been almost totally silent on this, the greatest moral and ethical challenge the human race has ever faced. I confess that I have not spoken out nor acted upon this great planetary crisis as frequently or with as much conviction as I should when I became aware of this catastrophe in the making. I have been timid when I have known the truth of the dire circumstances of our ecological crisis and that this is God's world, not ours, and we are called to care for it all. To care for life is the human species' most fundamental vocation (see Genesis 2:15), and we are failing to do so. In fact, with anywhere from 20,000-60,000 species of life going extinct every year, we are doing precisely the opposite; we are "de-creating" Earth. Why are we preachers silent? Why are we not raising this as a moral and spiritual affront to life, not just one Sunday a year around Earth Day, but with great frequency?

I am well aware that some of my colleagues consider addressing climate change as a luxury for those in communities that are not facing massive poverty, violence, lack of quality education, institutional racism, along with other acts of injustice. That is a valid critique. And yet over 50 years ago, leaders in my denomination, the American Baptist Churches, USA, reflected on this and began using the term "eco-justice" to point to the interconnection between ecological concern and human justice: when human beings are oppressed, Earth and her life systems are degraded; when Earth is abused, human beings, particularly the weakest and most vulnerable, suffer most profoundly. This happens all over the planet and in the United States, environmental racism is seen when "natural disasters" occur and it is the poor who suffer most profoundly, when land-fills are sited primarily in communities of color, when companies locate industrial plants in poor neighbors and call it economic opportunity, when the rates of cancer, neurological diseases, asthma and other environmentally related diseases are much higher in poor and non-white communities.

I'm a pastor. I'm tired of grieving with and burying more and more people who have died of cancer or neurological diseases. I am tired of burying those from the lower economic classes and people of color of diseases we should not be contracting in such massive amounts in the first place. And I am sickened that the first and most profound effects of climate change are being felt by those people who have the least to do with the problem in the first place! Maybe this time we'll wake up as a nation and demand that the fossil fuel industry stop their multi-billion dollar climate denial campaign now that the super storm destroyed rich, middle-class and poor, alike. We sure didn't wake up much when the poor and people of color where dying in New Orleans after Katrina, but maybe this time.... This is a matter of justice, a matter of caring for the "least of these, [Christ's] brothers and sisters."

God's first covenant followed the Flood (Genesis 9:8-17), a covenant between God, human beings and "all the creatures that came out of the ark." God's covenant was and is with all life, a covenant God will always keep. Have we? Where is the outrage for what we are doing to God's creation? Why have we remained virtually silent?

I am urging all of us to speak about the moral, ethical and spiritual reasons for addressing climate change.

I have spoken with many pastors and rabbis over the years who tell me that they don't touch climate change, because it is controversial. It has become too political, I am told, and if ... climate change was discussed in their congregation, people would leave. It's just too divisive, they tell me; not safe at all. I understand that. I am pastor of a diverse church which includes a few climate change deniers and other folk don't want to hear about it, and every time I raise this in worship or education or discussion groups, there are people who get upset with me and wish I would just shut-up and talk about "spiritual things." But I remember that Jesus was crucified because his teaching and life were too controversial for the religious, political and economic leaders of his day. He spoke and lived the truth and he was killed for it. This is the One we follow.

I am urging all of us to speak about the moral, ethical and spiritual reasons for addressing climate change.... Speak from your heart, from your faith that calls us to love what God loves, the world. If you don't feel adequate to speak on this, look at resources from the Evangelical Environmental Network (www.creationcare.org) or the National Religious Coalition on Creation Care (www.nrccc.org) or, if you are in Connecticut, contact the Interreligious Eco-Justice Network of Connecticut (www.irejn.org).

Jacques Cousteau once said, "We save what we love." Let's work to save the world that God so loves.

The Moral Imperative To Confront Climate Change

by Jeff Howard

The Unitarian Society of Hartford,
Hartford, Connecticut
November 3, 2013

Preface

The Green Sanctuary group at the Unitarian Society of Hartford (USH) sponsored a Sunday morning service dedicated to climate change. This homily was one of the elements of the service. Two explanations will help clarify references in the first paragraph: First, prior to the homily, members of the church's youth group who had slept in boxes on the church lawn to raise money for the homeless community spoke briefly about their experience. Second, throughout the service a series of more than 100 photographs flowed across a large TV monitor. They depicted homes damaged by Superstorm Sandy, glaciers melting, farmland parched, forests afire — as well as brilliantly colored autumn leaves, farmers markets, children playing, birds in flight.

The image of our young people spending the night in cardboard boxes is a bit disturbing, isn't it? The image of any young people sleeping in boxes is jarring. The image of young people sleeping in boxes during the long night is one we could have added to the set of images playing on the screen beside me. It is a fitting one for this particular service. For just as the flip-side of wealth is poverty, the flip-side of our fossil fuel economy is global impoverishment — impoverishment of families, impoverishment of cities, impoverishment of peoples, impoverishment of ecosystems, impoverishment of the oceans, impoverishment of life on this planet.

My name is Jeff Howard. For two years now I've been a member of the Green Sanctuary group here at the Unitarian Society of Hartford (USH). I have somewhat dreaded speaking here today, because I knew that even here, among friends, it would be painful to speak aloud, in a public place, about matters so painful to contemplate.

Climate change is real. It is a consequence of how we have built our society and fueled our economy. The threat it poses has grown more immense month after month, year after year....

Climate change is real. It is a consequence of how we have built our society and fueled our economy. The threat it poses has grown more immense month after month, year after year, decade after decade. Meanwhile, public cognizance of the magnitude of this threat is lagging dangerously.

You may have heard about an ignominious milestone earlier this year when the atmospheric concentration of carbon dioxide, climbing inexorably since the beginning of the Industrial Revolution, edged up to 400 parts per million. It reached 400 parts per million for the first time in human history. The last time the CO2 level was this high was 3 million years ago. Three million years ago! At that time, the Isthmus of Panama was forming and one of the prevailing hominids in Africa was Australopithicus.

To begin to grasp what 400 parts per million of CO2 in the global atmosphere means, consider sea level. Public discussion about sea-level rise almost always fixates on the 3-6 feet of rise expected this century. A rise of 3-6 feet in sea level will bring tides once a month that are equivalent to the storm surge that Superstorm Sandy produced; and it likely will spell calamity for Bangladesh and other low-lying, highly populous nations.

I see no honest way to avoid the conclusion that climate change is intimately related to deep defects in the way we have defined prosperity; the way we have conceptualized standard of living.... Climate change is the result of systemic flaws in our civilizational intelligence, our civilizational values, our civilizational self-conception.

But 3-6 feet is just the beginning of the story, my friends. The last time Earth saw 400 parts per million of CO2, the seas were not merely 3-6 feet higher. They were as much as 65 feet higher — 65 feet! Beyond the end of this century, and over the course of a number of centuries, if atmospheric carbon dioxide remains at today's level, we have every reason to expect sea levels high enough to drown thousands of square miles of Florida. High enough to bring the ocean to the doorstep of the White House. High enough to turn the modest Connecticut River into a vast estuary that, at Hartford, would be some 3 miles wide. High enough, perhaps, for the ocean to be lapping at the lower, eastern edge of what is now USH's own parking lot. If sea level rise is a key indicator by which to gauge the implications of our greenhouse gas emissions, the standard must not merely be 3-6 feet this century. It must include the realization that we've already put enough carbon in the atmosphere to utterly remake the coastline our heirs will inherit in the centuries that follow.

Dire though that scenario might seem, realize that it is based on an unrealistic assumption: that CO2 levels won't climb past 400 parts per million. No policy now in place, no trajectory of fossil fuel production now visible, no emissions scenario explored in climate models at MIT or NASA predict that CO2 levels won't continue rising well past 400 parts per million. The most optimistic scenarios have them rising to 450 or 500; others envision 700, 800, or higher. And the dynamics of atmospheric carbon — not to mention the dynamics of our economy and our politics — are such that getting back below 400 is unlikely in the foreseeable future, if ever.

Sixty-five feet of sea level rise is part of the historical vision that climate change forces upon us. Another part of that historical vision, equally important, is looking back — looking back to understand how climate change is rooted in the structure of our civilization and the structure of our own minds.

I see no honest way to avoid the conclusion that climate change is intimately related to deep defects in the way we have defined prosperity; the way we have conceptualized standard of living; the way we have understood power; the way we have thought about accomplishment and legacy;

the way we have framed development and growth; and, perhaps above all, the way we have conceptualized capital and profit. Climate change is not merely the result of mistakes, inadvertence, good intentions gone awry. Climate change is the result of systemic flaws in our civilizational intelligence, our civilizational values, our civilizational self-conception.

The flip side of our idea of The Good Life, it turns out, is killer heat waves, freak blizzards, and melting of the Arctic ice cap. The flip side of our idea of control is oceans becoming acidic and rising higher than they have been in all of human experience. The flip side of our idea of wealth is the collapse of ecosystems, along with drought and famine on an unprecedented scale. The flip side of our idea of investment is killer superstorms and tropical diseases moving into what we have thought of as temperate latitudes. The flip side of our idea of energy is millions of "climate refugees" fleeing the sea, fleeing famine, fleeing social disarray. The flip side of our idea of growth is extinction of many species and methane bubbling up from lake beds and sea floor.

Recalibrating, rethinking, reimagining, reinhabiting these ideas — here lies hope. Revising, reforming, revolutionizing the material enactment of these ideas — here lies hope. Taking responsibility for leading this process — here lies a moral response to climate change.

We will not get out of this predicament quickly or easily. It will take generations of struggle. It will take generations of struggle, with each step as painful as ending slavery in the American South, each step as challenging as ending apartheid in South Africa, each step as tumultuous as ending the Vietnam war and securing the right of blacks and women to vote and own property. Taking responsibility for leading this process — here lies a moral response to climate change.

On Thursday of this week, scores of people from a wide range of religious traditions will gather in Hartford for what is being billed as the Climate Stewardship Summit. The idea for the event was spawned in this very building. In March of this year, the Green Sanctuary group hosted a meeting of people from the Hartford area who are concerned about climate and ready to act. About half of us at the meeting resolved to mount a statewide summit. Organizing is being done by an interfaith nonprofit organization, the Interreligious Eco-Justice Network of Connecticut (IREJN), with a steering committee consisting of members of this congregation and numerous others. The Green Sanctuary group is one of about a dozen organizations sponsoring the summit. The theme of the event is "Climate action as a moral imperative for Connecticut communities of faith."

Climate change is happening too fast — and our civilization's engagement of the problem has been too slow — for the world's people and ecosystems to avoid pain and suffering on a very large scale.

Across the state and across the country, religious communities of many shapes and sizes — UU, Episcopal, Baptist, Catholic, Muslim, Quaker, Jewish, and others — are stepping forward to lead. They're installing photovoltaic systems and geothermal systems, and they're buying electricity from wind farms. They're upgrading their buildings to make them more efficient. They're publishing op-ed pieces in local newspapers confronting climate change denialists. They're setting up carpools and putting biodiesel in the tanks of their buses. They're holding special services on climate change, like this one. They're challenging their members to do household energy audits, use mass transit, and cut down on meat consumption. They're designing Sunday School lessons around climate change. They're sending delegations to statehouses to lobby for clean-energy laws. They're building up their food pantries and crisis-response teams to prepare for the next superstorm. They're holding climate change summits and adopting statements of conscience. And they're participating in a nationwide movement to purge fossil fuel industry investments from their endowment portfolios — an idea that the Green Sanctuary group is preparing to broach with USH's Endowment Committee. In short, across the country, even those who are not UUs are, in effect, following the UU principle "service is our prayer." Here lies hope. Here lies a moral response to climate change.

> *I'm heartened by the prospect that we can harness this painful experience to make a better civilization — a citizenry more clear-eyed and more fully human; governments and economies and businesses more humane and just....*

Climate change is happening too fast — and our civilization's engagement of the problem has been too slow — for the world's people and ecosystems to avoid pain and suffering on a very large scale. But I'm heartened because I know that even now, we can significantly blunt the damage and suffering. I'm heartened because I know that even at this late date, we can still bring the politically powerful fossil fuel industry to heel. I'm heartened because I know that even at this eleventh hour, with colossal changes in natural systems already set in motion, we can learn from our mistakes and begin making the changes in human systems that will allow us to stop compounding the damage and misery. I'm heartened because I know that we can still learn to inhabit this blue planet sustainably.

I'm also heartened by the prospect that we can harness this painful experience to make a better civilization — a citizenry more clear-eyed and more fully human; governments and economies and businesses more humane and just. This better civilization won't magically appear; we will have to coax it out of a dire situation. We must make the deeply humbling, tragic experience of climate change count for something essential.

May it be so.

_____

End Notes

"Rising seas," National Geographic, Sept. 2013
http://ngm.nationalgeographic.com/2013/09/rising-seas/folger-text

"When crocodiles roamed the poles," CNN, June 25, 2013
http://www.cnn.com/2013/06/25/opinion/pagani-climate-change/index.html
Map depicting various levels of sea level rise (for 65' of rise, select +20 meters from the menu at top left) http://flood.firetree.net/

Climate Stewardship Summit, Hartford, CT, Nov. 3, 2013
http://www.irejn.org/what-we-do/climate-stewardship-summit/

NATIONAL CATHOLIC REPORTER

Editorial: Climate Change is Church's No. 1 Pro-life Issue

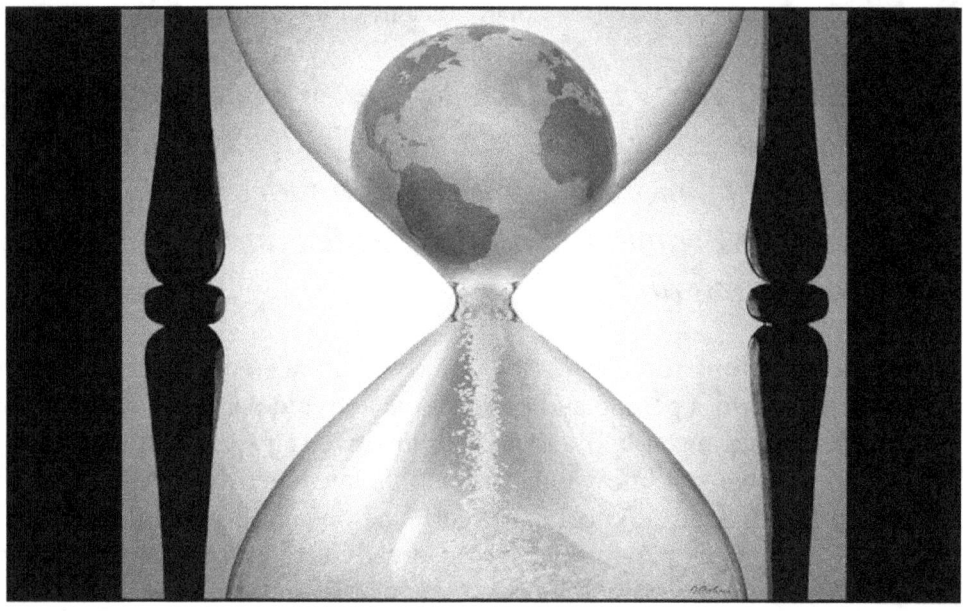

NCR Editorial Staff | May. 20, 2014

THERE MAY HAVE BEEN A TIME when moving from a point of indecision on the matter of climate change, to a decision on whether it is real and caused by humans or not, required leaps of faith of somewhat equal proportions. But that was a long time and a lot of science ago.

The science, as it has developed, may not be perfect, but it is long past time that the question turn from whether human activity is causing climate change to what do we do about it. The Catholic Church should become a major player in educating the public to the scientific data and in motivating people to act for change.

The case for the reality of human-caused climate change was made in the strongest terms to date in the recently released third National Climate Assessment, a report exhaustive in its detail and the manner of its preparation. It was compiled by a team of more than 300 experts, including policymakers, decision-makers from the public and private realms, researchers, representatives of business and nongovernmental organizations, as well as representatives of the general public.

It was reviewed extensively, including by a panel of the National Academy of Sciences, the 13 federal agencies of the U.S. Global Change Research Program, and the federal Committee on Environment, Natural Resources, and Sustainability.

As columnist Michael Gerson wrote of suspicions that the rising awareness of climate change was a product of scientific fraud: "In this case the conspiracy would need to encompass the national academies of more than two dozen countries, including the United States." The larger point he makes is that we need to get on with the questions that only science can address.

The National Climate Assessment report doesn't speak of the crisis as a moment to anticipate but as increasingly evident today. The most frightening point is that climate change "is projected to continue, and it will accelerate significantly if global emissions of heat-trapping gases continue to increase." Those changes, in turn, will increasingly threaten humans "through more extreme weather events and wildfire, decreased air quality, and diseases transmitted by insects, food, and water."

In an opportune coincidence, about the same time the U.S. government was releasing the National Climate Assessment, the Vatican was releasing a far shorter summary of its five-day summit on "Sustainable Humanity, Sustainable Nature: Our Responsibility."

While the church has taken it on the chin for centuries-old condemnations of scientific truths, the reality today is that it stands uniquely in a position to not only aid the science but also to engage in the ethical discussions essential to any consideration of global warming.

If there is a certain wisdom in the pro-life assertion that other rights become meaningless if the right to life is not upheld, then it is reasonable to assert that the right to life has little meaning if the earth is destroyed to the point where life becomes unsustainable.

Cardinal Oscar Andres Rodríguez Maradiaga described the problem during a talk opening the Vatican conference. He described nature as neither separate from nor against humanity, but rather existing with humans. "No sin is more heartless than our blindness to the value of all that surrounds us and our persistence in using it at the wrong time and abusing it at all times." Humans, he said, have become technological giants, while remaining ethical children.

Humans have been driven to a point of decision by the consequences – good and bad – of two centuries of technological development. In his closing remarks at the Rome meeting, *The New York Times* writer Andrew C. Revkin stated, "Scientific knowledge reveals options. Values determine choices."

"That is why the Roman Catholic church – with its global reach, the ethical framework in its social justice teachings and, as with all great religions, the ability to reach hearts as well as minds – can play a valuable role in this consequential century."

The problem is enormous, but so is the opportunity for the church to use its resources, its access to some of the best experts in its academies and the attention of those in its parochial structures to begin to educate. This is a human life issue of enormous proportions, and one in which the young should be fully engaged. The Climate Assessment document as well as the recent discussion at the Vatican are excellent starting points for developing curricula materials for education programs in parishes and schools.

Catholic high schools and colleges have the freedom to explore these vital issues from both the scientific and ethical perspectives. They can bring theological perspectives to bear on the issues. Educators and students could devise ways to become active at all levels, from homes, to communities, to states, to advocating for legal measures to offset the effects of global warming.

Finding a fix for climate change and its potentially disastrous consequences, particularly for the global poor, is not the work of a single discipline or a single group, or a single political strategy. Its solution lies as much in people of faith as in scientific data, as much or more in a love for God's creation as it does in our instinct for self-preservation.

The Meaning of This Hour

Confronting the Coming Cataclysm of Global Climate Change

by Rabbi Larry Troster
Teaneck, New Jersey
October, 2013

In March 1938, Abraham Joshua Heschel delivered a speech to a conference of Quakers in Frankfort (it was later expanded and published in 1943) called *The Meaning of this Hour*.

Heschel had been living in Berlin for some years, acquiring his Ph.D. and a liberal rabbinic ordination (he had already gotten a traditional ordination when he was a teenager in Warsaw). During his years there, he was a witness to rise of Nazism even while he taught and began to publish his work.

In 1938, it was clear to many people that war in Europe was coming. In the very month that Heschel spoke came the *Anschluss*, the Nazi takeover of Austria. Heschel was arrested in October of 1938 and deported to Poland. Six weeks before the invasion of Poland in September 1939, Heschel was able to get to England and from there to the United States. In a speech given in 1965 called *No Religion is an Island*, he referred to himself as "a brand plucked from the fire in which my people was burned to death." (He was alluding to Zechariah, chapter 3, where the High Priest Joshua, who had been born during the exile in Babylon and was one of the first to return to Judea, was called by God, "a brand plucked from the fire.")

> *Climate change is one of the greatest moral disasters of human history as the people who will suffer the most have been the least responsible for its cause.*

Heschel warned of the coming cataclysm in vivid and forceful language, evoking images of the demonic. He said, "At no time has the earth been so soaked with blood. Fellowmen turned out to be evil ghosts, monstrous and weird." He asked the question, "Who is responsible?" We are, he said, by not fighting for "right, for justice, for goodness." He said that we should be ashamed, and after the war, when the full horror of the Holocaust was revealed, he said that we should not ask, "Where was God?" but "Where was man?"

While we are not facing another world war and I am usually loath to reference the Holocaust when dealing with contemporary issues, I could not but be struck by the urgency of Heschel's speech when I think about the looming disaster of climate change.

The meaning of *this* hour is that we are continuing to argue about the fact of climate change when it is already happening and millions of people are already feeling its effects. Droughts, floods, increases in forest fires, seas rising and thousands of scientific indicators seem not to move us. Several years ago, CARE published a report on climate refugees which *conservatively* estimated that by 2050 there would be 250 million climate change refugees. A long lasting drought in the

Middle East was one of the factors which precipitated the civil war in Syria, just one more example of how climate change has and will cause unrest, strife and war.

The Intergovernmental Panel on Climate Change (IPCC) is a coalition of thousands of scientists worldwide who have been tracking and evaluating the research on climate change since 1988. Its fifth assessment report will be issued later this month. A draft of that report was leaked to reporters last month and it says that it is "extremely likely" that human actions are the cause of most of the temperature increases of the last sixty years. "Extremely likely" is the way scientists say something is 99% certain. They wrote, "There is high confidence that this has warmed the ocean, melted snow and ice, raised the global mean sea level and changed some climate extremes in the second half of the 20th century." And things could get much worse. If carbon dioxide and other greenhouse gases continue to be emitted into the atmosphere at present rates, global temperatures will rise by more than 5 degrees Fahrenheit. This would cause large scale melting of ice, more extreme heat waves and flooding, disruptions in the world food supply and the massive extinction of plant and animals species.

The IPCC, because it is a collective group of scientists under the auspices of the United Nations, has always been conservative in its assessments. Many climate scientists believe that the situation is even worse and some believe that we may be too late to avoid a catastrophic change in the global climate. To some extent they are right. Even if we were to eliminate all the carbon emissions today, the CO2 already in the atmosphere will continue to have an effect for hundreds of years. But we can stop this situation from getting more dangerous.

In the published version of his speech Heschel wrote, "The Almighty has not created the universe that we may have opportunities to satisfy our greed, envy and ambition. We have not survived that we may waste our years in vulgar vanities." These words can easily apply to our lack of action on climate change. We often think that it is all a matter of technology; that we can somehow come up with some gadget that will make all the CO2 go away without our having to change anything about the way we live. The only way to prevent a disaster for future generations is to phase out carbon based energy as quickly as possible. And to do that, we need to act now.

In the Haphtarah for Yom Kippur morning, we read Isaiah 57:14 - 58:14. In this passage the prophet says that people don't understand why God has not forgiven them even though they have fasted. God replies that their ritual is hypocritical because even while they fasted they have acted immorally by oppressing their workers. A true fast, says God, must be one that accompanies justice and the care of the poor and powerless. Only then, will God answer, *Here I am*, when you call.

Climate change is one of the greatest moral disasters of human history as the people who will suffer the most have been the least responsible for its cause. Those of us in the developed countries somehow think that we will escape its results, turning away from the hundreds of millions who will be caught in the whirlwind of misery that is coming.

The meaning of *this* hour is that we must recognize what we are doing, admit our fault and bring about the necessary changes to prevent further damage. Fifty years ago, Martin Luther King Jr. spoke of the "fierce urgency of Now." Once again, *that* is the meaning of this hour.

(This was originally published in the <u>New Jersey Jewish Standard</u>)

Revelation: Playing God Is Bad for Business

by the Reverend Charles Redfern
September 9, 2014
Reprinted from *The Huffington POST*

IT'S A STRANGE HARMONY. WALL STREET POWERHOUSES now chime with robed clerics in an offbeat tune: Ruining our atmosphere is not only immoral, it's bad for the S&P 500.

What's next? Will the executives and clergy swap portfolio tips for divestment brochures at the People's Climate March on September 21 in New York City?

It could happen. Just listen to the chimes.

The first rings from a study published by the bipartisan "Risky Business Project," whose co-chairs hardly conjure images of radical vegans: former New York Mayor Michael Bloomberg; former U.S. Treasury Secretary Hank Paulson, who served under President George W. Bush; and Tom Steyer, a retired hedge fund manager. Robert Rubin, President Clinton's Treasury Secretary and a "risk team" member, explained the project's conclusions in a *Washington Post* op-ed: Some frame the climate debate as "a trade-off between environmental protection and economic prosperity," with carbon reduction poised as the spiteful job killer. "That's precisely the wrong way to look at it," he said. "The real question should be: What is the cost of inaction?"

Rubin's answer: Immense. By 2050, for example, between $48 billion and $68 billion in Louisiana and Florida property will lie below sea level – which doesn't exactly pave the path for robust real estate values and a healthy tax base.

> *Our civilization now faces an unavoidable option:*
> *Will we choose the difficult path that leads to hope?*
> *Or will we cater to wistful, hear-no-evil dreams*
> *from which we'll awaken into a bleak dystopia?*

Three cheers for Rubin and his colleagues. They're peering beyond the latest quarterly report – and they're finally in tune with the other orchestra, the one that plays on the field of spirituality and morality. Its chime rings: A withered spirit is even more alarming than a dropping GNP. All major monotheistic faiths say God vested the Earth into human care. We're here to nurture creation as divinity's representatives, not destroy it. We sing with the psalmist: *"The earth is the Lord's and all that is in it, the world, and those who live in it; for he has founded it on the seas, and established it on the rivers"* (Psalm 124:1).

We've been singing that verse for decades. Marybeth Lorbiecki documents the environmental legacy of Pope John Paul II in *Following St. Francis: John Paul II's Ecological Call*

To Action. He was frank in 1990: "In our day, there is a growing awareness that world peace is threatened not only by the arms race, regional conflicts and continued injustices among peoples and nations, but also by a lack of due respect for nature, by the plundering of natural resources and by a progressive decline in the quality of life. The sense of precariousness and insecurity that such a situation engenders is a seedbed for collective selfishness, disregard for others and dishonesty."

He joined Eastern Orthodox Patriarch Bartholomew I in 2002 and issued a Common Declaration on Environmental Ethics: "The problem is not simply economic and technological; it is moral and spiritual." They called for a "new approach" and a "new culture."

In other words, the current Pope Francis was saying nothing new in 2013 when he pleaded to world leaders: "Safeguard Creation, because if we destroy Creation, Creation will destroy us! Never forget this!"

Similar clarions ring from The Central Conference of American Rabbis, who called for "strong action," as well as dozens of Evangelical Christian leaders: "The need to act now is urgent." The interfaith National Religious Coalition on Creation Care was especially blunt in a recent statement (full disclosure: Roman Catholic theologian Richard W. Miller and I wrote the first of its many drafts): "Our civilization now faces an unavoidable option: Will we choose the difficult path that leads to hope? Or will we cater to wistful, hear-no-evil dreams from which we'll awaken into a bleak dystopia? The difficult but rewarding path brings us to a richer vision of ourselves, our Creator, and creation. We'll either re-forge the bonds between spirituality, morality, and policy or our dreams for the future will abandon us on the rubble of a modern-day Tower of Babel."

Coming to grips with the environmental crisis may rekindle a deeper vision of who we are and where we live.

Perhaps we've been numbed. Perhaps the dire scientific warnings and the oft-repeated "97 percent" figure have congealed into white noise (97 percent of all climatologists agree that human-induced climate change is a scientific fact). Maybe we even skipped a recent Associated Press story: The UN Intergovernmental Panel on Climate Change sent international leaders a final draft of its "synthesis report," which fuses three earlier, heftier documents into an easier-to-read 127-page summary. Its message: "currently observed impacts might already be considered dangerous."

Maybe the new harmony will ring above the noise and awaken us to a long-buried reality: Humanity didn't view itself as a mere economic creature until the so-call Enlightenment - which, as Alister McGrath argues in *The Re-Enchantment of Nature*, not only brought laudable scientific and technological advances. It also hammered the first planks in the philosophical platforms supporting Gulags and Holocausts. Coming to grips with the environmental crisis may rekindle a deeper vision of who we are and where we live.

So I hope. Otherwise, the chimes may transform into death knells.

The Malankara Church of South India

Why Climate Change is an Issue of Justice

Rev. Dr. Royce M. Victor
The Church of South India
Oikos News, June, 2014

ONE OF THE MAJOR ISSUES OUR PLANET FACES TODAY is ecological crisis, more specifically, climate change, which above all is a justice issue. The people who are already suffering most from climate change/global warming are those who have done the least to cause it, and have the least resource to do anything about it. For example, the nomadic tribes in northern Kenya, who walk hundreds of kilometers in search of water, only to encounter danger and conflict when they find some. The poor farming families in Bangladesh, forced by rising sea levels to leave their homes for temporary and illegal camps on embankments, where they cannot earn a living. None of these people have more than a minimal responsibility for CO2 emission that have caused, and continue to exacerbate, the current crisis.

Also, whenever there is an ecological crisis, whether in a developed country or in a developingcountry, the poor suffer the most. Their lives are always at risk. The people who have money or power or influence escape easily from it. On the one hand, the rich nations have reached a peak of development, on the other; the poor nations are still struggling to get on to the development ladder. So the justice issue of climate change is directly related to the development model of today's world.

As Desmond Tutu points out, 'just as we are all part of the problem, so we are all also part of the solution. We all need to come to terms with the forces that have created this crisis and the resources within our traditions that can motivate us to resolve the crisis. One of those traditions is our biblical heritage.' These words clearly elucidate the imperative to look at the bible for guidance, inspiration, and encouragement to resolve the crisis.

Climate change is not mentioned in the Bible directly, nor has it so far been part of contemporary systematic theology. Very recently Christian theologians started thinking seriously about biblical reflection on climate change. Now increasing number of theologians and Christian believers are of the opinion that the God of the Bible is a God of justice who protects, loves and cares for the most vulnerable among God's creations.

They also find that the present world development model is threatening the lives and livelihoods of many, especially among the world's poorest people, and destroying biodiversity. Therefore, they affirm that the Christian vision should be aimed at the possibilities to overcome/dethrone the existent model of development based on over-consumption and greed.

The people who are already suffering the most from climate change/global warming are those who have done the least to cause it, and have the least resource to do anything about it.

Recently, Pope Francis made the religious case for tackling climate change, calling on his fellow Christians to become "Custodians of Creation" and issuing a dire warning about the potentially catastrophic effects of global climate change. He argues that respect for the "beauty of nature and the grandeur of the cosmos" is a Christian value, and failure to care for the planet risks apocalyptic consequences. He indicates, "Because if we destroy Creation, Creation will destroy us! Never forget this."

He further points out, "Creation is not a property, which we can rule over at will; or, even less, is the property of only a few: Creation is a gift, it is a wonderful gift that God has given us, so that we care for it and we use it for the benefit of all, always with great respect and gratitude." Pope Francis also says that humanity's destruction of the planet is a sinful act, likening it to self-idolatry. "When we exploit Creation we destroy the sign of God's love for us, in destroying Creation we are saying to God: 'I don't like it! This is not good!' 'So what do you like?' 'I like myself!' – Here, this is sin! Do you see?" Throughout the speech he reiterates that environmental justice and economic justice are inextricably connected to each other.

Southern African Faith Communities

Southern African Faith Leaders Speak Out on Climate Change

Introducton

http://safcei.org/wp-content/uploads/2011/07/SA-Faith-Leaders-Declaration-09-05-2011.pdf

This Statement was issued by representatives of religious organisations from across Southern Africa gathered in Lusaka, Zambia, from 5-6 May 2011 to discuss and prepare the response of faith communities to climate change and our response to the UN climate change negotiation in Durban, COP-17.

In the upcoming climate negotiations, South Africa must stand with Africa and its people – not with the big polluters.

Africa is a continent of the faithful. God has entrusted us with a rich, living planet. Protecting the environment is demanded of us by our faiths.

It is clear that Africans are already profoundly and directly affected by climate change. We faith leaders of Southern Africa wish to remind South Africa that it is part of Africa, and that we stood with South Africa during the dark years of apartheid. We now call on South Africa to stand again with its neighbours rather than continuing with a growth path that will help make all our countries victims of a climate change disaster.

Faith leadership and climate change

We, the people of the world, have lost our moral compass, and reduce all economic decisions to maximising profit and consumption, and so as faith communities we must renew our commitment to compassion for other living beings and the principle of justice.

We note that climate change is a systemic crisis of an unsustainable economic model, and without substantial changes to that system, without establishing an alternative sustainable economic model, we have little chance of averting the worst effects of climate change.

Ultimately, it is the human desire to live lives filled with love and peace that drives all of our other desires. Therefore, let us restore justice, love, and a love for peace, and bring these qualities to the heart of our climate negotiations.

South Africa must show true leadership for sustainability

As faith leaders from southern Africa, we believe that at this "African COP" to be held in Durban in November. The countries of Africa must take a lead and set an example by seeking climate justice, reducing carbon emissions and avoiding further emissions increases.

As the host country, South Africa should set an example to the world by committing to drastic and radical reductions, without waiting for global climate funding or commitments by other countries. We believe that we in Africa, and specifically South Africa as the host country, must take a moral stand and challenge the world to "do the right thing," to break the logjam and bring about a meaningful and legally binding agreement.

For South Africa, true leadership on climate change and sustainability must mean abandoning nuclear energy and its continued use of coal, its insistence on claiming further carbon space, and its refusal to change unless it is paid to do so by the international community. It must turn away from supply and pricing models that privilege multinational corporations, must improve on its current paltry ambition of at most 20% renewable energy by 2030, and commit resources into developing renewable energy and promoting energy efficiency that will create new employment and new opportunities for Southern Africa.

We commit ourselves to action:

- We will set a good example in our personal lives by reducing all forms of over-consumption.

- We will lead our faith communities and wider communities to understand the threat of climate change and the need to build alternative economies and societies based on deeper values.

- We will build relationships with global faith communities in pursuit of our common goals and objectives.

- We acknowledge that climate change has disproportionately affected women, and that it worsens existing inequalities – therefore, addressing these inequalities is ever more urgent.

- We commit ourselves to finding ways of supporting our African negotiators who are currently few in number, too often replaced and usually terribly under-resourced.

- We will support the struggles of people affected by regional large mining projects.

We call on world leaders to:

- ◆ Commit to the principle of inter-generational equity and the rights of our children.

- ◆ Commit to understanding and establishing the rights of Mother Earth as outlined in the Cochabamba declaration.

- ◆ To abandon GDP as an indicator of economic wellbeing in favour of indicators that include human wellbeing, equality and the external environmental costs of human economies.

- ◆ To set final targets for phasing out the use of all fossil fuels, and deep interim reductions in carbon emissions that support the target of no more than one degree of global warming.

- ◆ To ensure that there is sufficient climate finance for adaptation in Africa, additional to existing development aid

- ◆ To channel sufficient climate financing from the historic polluting nations in recognition of their ecological debt to enable Africa to leapfrog into an age of clean energy technology.

Anglican Archbishop Rowan Williams warns of impending climate catastrophe

The former Archbishop of Canterbury says that Western lifestyles bear responsibility for causing climate change in the world's poorest regions

Dr. Rowan Williams' plea to combat global warming by reducing consumption of fossil fuels comes on the eve of the most authoritative study yet into the impact of climate change.

By Robert Mendick, Chief reporter
The London Telegraph
29 March 2014

Dr Rowan Williams, the former Archbishop of Canterbury, has attacked Western lifestyles for causing climate change that is "pushing the environment towards crisis."

Writing in the Telegraph, Dr Williams says that the "appalling" floods and storms that devastated parts of Britain this winter were a demonstration of "what we can expect" in the future. The floods in Britain and weather-related "catastrophes" in the poorest countries on Earth, he insists, are the clearest indications yet that predictions of "accelerated warming of the Earth" caused by "the uncontrolled burning of fossil fuels … are coming true."

On Monday, the UN's Intergovernmental Panel on Climate Change (IPCC) will publish its latest study… on the consequences of the predicted rise in global temperatures. The report… will claim the cost of combating the effects of a 4.5 F (2.5 C) rise in temperature by the end of the century will be £60 billion a year. It will warn that… climate change will be felt most keenly in Africa, South America and Asia and predicts droughts, food shortages and a rise in diseases such as malaria.

Dr. Williams writes: "We have heard for years the predictions that the uncontrolled burning of fossil fuels will lead to an accelerated warming of the Earth. What is now happening indicates that these predictions are coming true; our actions have had consequences that are deeply threatening for many of the poorest communities in the world.

"Rich, industrialised countries, including our own, have unquestionably contributed most to atmospheric pollution. Both our present lifestyle and the industrial history of how we created such possibilities for ourselves have to bear the responsibility for pushing the environment in which we live towards crisis."

Dr. Williams, writing in his capacity as chairman of Christian Aid, said that the winter storms that battered Britain had brought climate change to the fore in this country and that the IPCC report, published at a specially convened meeting in Yokohama, Japan, puts "our local problems into a deeply disturbing global context."

> *We have heard... predictions that burning of fossil fuels will lead to warming of the Earth. What is now happening indicates that these predictions are coming true....*

The IPCC, he says, will be "pointing out that ... we [the UK] have in fact got off relatively lightly in comparison with others." While the "chaos [of the flood] came as a shock to many," other countries such as Bangladesh and Kenya, among others, have suffered far worse catastrophes caused by climate change....

Dr. Williams goes on to attack global warming sceptics and climate change deniers. "There are some who doubt the role of human agency in creating and responding to climate change, and who argue that we should direct our efforts solely to adapting to changes that are inevitable, rather than modifying our behavior.... That approach might be all very well in the UK where flood defenses and other measures can be adopted relatively cheaply, but in the most vulnerable, poorest countries worst affected by global warming that is not an option."

Dr. Williams's intervention in the climate change debate comes as officials and researchers in Japan finalise the IPCC study. The report...will focus on the impacts of global warming. It is expected to say that Africa will be affected by longer droughts that threaten livestock and crop yields. The IPCC expects to see worsening health as a result with increased malnutrition, malaria and other diseases. Rising temperatures will also affect food production and security in parts of Asia with a fall in rice yields caused by a shorter growing period.

The IPCC report will say that northern parts of Asia may benefit from warmer temperatures, leading to increased production of wheat and other cereals. In South America, ice and glaciers in the Andes are "retreating at an alarming rate," affecting water supplies while "unique ecosystems" are threatened by climate change and increasing industrialisation.

Why Pro-Life Christians Are Addressing Climate Change

by Reverend Joel Hunter
Reverend Mitch Hescox and
Alexei Laushkin

May 13, 2014

> From the formation of a child's first tiny cell to life's final breath, all life has dignity and value because each and every one of us is made in the image of God. And that is why when we talk about being "pro-life," it's not just about a political issue. It's a worldview…it's a life-view. It's a way of looking at each human life that transcends culture, class, race, age and opinion.
>
> — *The Dignity of Life,* by Focus on The Family

WE BELIEVE THAT CREATION CARE IS A MATTER OF LIFE because we see a clear scriptural ethic to protect human life at all stages; from conception to natural death. This view is anchored in historic Christian teaching on the subject and it is the same ethic that motivated early Christians to take up adoption and what motivates Christians in this age to protect the unborn from abortion. As the recent video, "The Dignity of Life," by Focus on the Family puts it: "From the formation of a child's first tiny cell to life's final breath, all life has dignity and value because each and every one of us is made in the image of God."

For us, being pro-life includes not only defending our unborn children, but also the biblical mandate to care for all life. While the threats may be different, the injunction to protect life is the same. We are called to protect this seamless garment of life.

> *We believe climate change to be a profound pro-life issue.… While it is good to respond to current challenges, it is even more cost effective to spend funds ahead of time to prepare for present changes in the climate, including extreme weather events.*

Toxins and other pollutants foul our water, air, and soil, impacting the purity of life God intends. Children are especially vulnerable to many of these pollutants because their small bodies are still developing. A few years ago pro-life evangelicals spoke out on the impact of mercury on the unborn. 1 in 6 children in the U.S. were born with too high levels of mercury in their blood; here's an audio briefing on why mercury is so dangerous for the unborn. Because of the efforts of pro-life evangelicals the United States is taking a leadership role in reducing the impact of mercury on the unborn. Another important issue is water. As a recent *USA Today* op-ed put it if you care about life pay attention to what's happening with water.

We believe climate change to be a profound pro-life issue, and Florida is ground zero when it comes to climate change. Cities across the state are already spending millions in taxpayer dollars to install new sea level pumps, bolster sea walls, and protect from salt water intrusion. While it is good to respond to current challenges, it is even more cost effective to spend funds ahead of time to prepare for present changes in the climate, including extreme weather events. Let's upgrade Florida's water pumps and building codes today before we have to clean up a bigger mess tomorrow. Given the dollars already being spent and scale of the cost, if you care about taxpayer money and limited government. you should care about climate change. We are also concerned about worsening air pollution under climate change. Duval County alone has almost 18,000 cases of pediatric asthma. That number would be dramatically lower if we were better stewards of God's world.

We need to see climate, not as an issue about politics or partisanship, but as a moral concern.... As the church starts to take on climate change more directly, it's also time for clean businesses to take the lead.

When we see the present impacts our pro-life ethic kicks in. Let's empower individuals to take the lead when it comes to entrepreneurial business solutions that create a cleaner environment. We need to see climate not as an issue about politics or partisanship, but as a moral concern. God has given us all the tools to be good stewards of God's creation. It's time for Florida to come together to come up with a plan to address climate change. The church in Florida is already starting to take the lead through "the Joseph Pledge." As the church starts to take on climate change more directly, it's also time for clean businesses to take the lead. The cost of solar has plummeted, yet Florida is still well behind where it could be when it comes to clean energy. We need to do what we can to transition away from expensive fossil fuels and toward cheaper and healthier technologies. These actions should include putting together a plan for Florida to play a part in achieving the Clean Power Plan and finding conservative solutions to addressing carbon pollution.

Every concern mentioned in the video by Focus on the Family is impacted by our poor stewardship of God's creation, whose consequences are borne by our children in their bodies and the future we bequeath to them. If creation isn't stewarded well how do we expect the poor to have access to fresh food and to live free of toxins in their neighborhood? Our poor stewardship of God's world is a reflection of how seriously we take God's teaching. That's why creation-care remains integral to being pro-life. As the Focus video states, being pro-life is "not just about a political issue. It's a world view – it's a life view."

Rev. Joel Hunter is Senior Pastor of Northland Church, a Church in Longwood, Florida; Rev. Mitch Hescox is the President of the Evangelical Environmental Network; and Alexei Laushkin is the Vice-President of the Evangelical Environmental Network.

"As in Heaven, So on Earth"

The Christian Principle that Addresses Climate Change

by Vincent Rossi
Saint Seraphim Orthodox Church (OCA)
Santa Rosa, California

TO SAY THAT SIN IS THE SPIRITUAL ROOT OF ALL HUMAN POLLUTION, as does the Ecumenical Patriarch and many other commentators, is implicitly to affirm three spiritual truths:

> (1) That heaven and earth, the Uncreated and the created, are reciprocally linked and interpenetrate one another; (2) that the transfiguration of the earth involves the continuous communion of heavenly and earthly realities through human right thinking, feeling and acting in response to the reality of the Sacred in all things; and (3) that the sundering of the link between heaven and earth, between God and creation, leads inevitably to the desacralization, desecration and destruction of creation.

The sundering in human consciousness of the link between heaven and the earth, that intimate interpenetration of the Divine Archetype and the visible image which is the source of the Sacred, is at the heart of the persistent and diabolical insensibility in human beings, even Orthodox Christians, to the desecration of creation.

The divorce of the created order from the Uncreated, which leads to a widespread belief that the purpose of Christ's coming is exclusively to save humanity by teaching us all to devalue and abandon the earth and concentrate solely on "getting to" heaven, is profoundly unbiblical and un-Orthodox. The "kingdom of heaven" of the Gospels is not a "place" disassociated from the earth, but is the Presence, Power, Energy, Mind, Grace and Will of God "on earth as it is in heaven," as the Lord's Prayer teaches us to pray.

The Lord's Prayer is an indispensable canonical element in the Divine Liturgy and in the entire Orthodox daily cycle of services, not merely because it is authored by the Lord Himself, but primarily because of its spiritual and theological centrality and universality. It places before all men universal and inescapable truths to be contemplated and prayed for. The theological center of the prayer, the axis of energy around which its seven petitions revolve, lies in the phrase "on earth as it is in heaven." Scripturally, patristically, liturgically, sacramentally and theologically, the aim of the Church as the living Body of Christ has always been to realize on earth the order, pattern and reality of heaven, not merely to escape the earth in order to "go to" heaven.

The implications of this truth its loss — are immense. It includes everything created, everything human, without restriction, without limit. There is nothing created that the genius of this prayer does not incorporate. We pray for the Father's Kingdom to come and for His Will to be done — not in vague or general terms — but precisely on earth, as it is in heaven.

The Lord's Prayer is the very charter requiring Orthodox Christians to resacralize the earth as part of our own salvation and theosis. How can we possibly say this prayer daily without being totally convicted for the weakness and half-heartedness of our response to the cry of the earth?

Why the Urgency on Climate Change? Scientific and Other Perspectives

SCIENCE PROVIDES THE EYES FOR SOCIETY WHILE RELIGION SERVES AS ITS heart and moral conscience. The consistent conclusion from peer-reviewed science is that climate change is the most urgent issue facing our generation.

Society has known about climate change for more than twenty years. Back in 1994, over a thousand top scientists issued a warning to humanity. They declared that humanity and the natural world are on a collision course. *"Human activities,"* they said, *"inflict harsh and often irreversible damage on the environment and on critical resources. If not checked, many of our current practices put at serious risk the future that we wish for human society.... Fundamental changes are urgent if we are to avoid the collision our present course will bring about."*

In the years since this warning, individuals and businesses have continued to increase atmospheric carbon dioxide and jeopardize climate stability with callous disdain for life's sanctity. Meanwhile human populations are growing and putting greater pressure on the planet's life support systems.

Thus the eyes of science may explain our predicament, but the moral conscience of religion, especially at the parish and congregation level, appears to be mindlessly cooperating with a progress and unsustainable growth agenda.

It is time to wake up. Thirty years ago, when Christians and Jews began to acknowledge responsibility for creation, they found in Scripture principles and teachings that could transform our rapacious impact upon the world. If clergy especially could listen to their own teachings, we would take action to address the evils of consumerism and the impact of fossil fuels upon the world. We would promote clean energy and teach the urgency of climate change that scientists are observing along with many others in society.

In the following articles, read how science and society leaders view the dangers of global climate change.

Climate Change Is Happening Now – A Carbon Price Must Follow

The extreme weather events of 2012 are what we have been warning of for 25 years, but the answer is plain to see

By Dr. James Hansen
Director, NASA
November 30, 2012

WILL OUR SHORT ATTENTION SPAN BE THE END OF US? Just a month after the second "storm of a century" in two years, the media moves on to the latest scandal with barely a retrospective glance at the implications of the extreme climate anomalies we have seen.

Hurricane Sandy was not just a storm. It was a stark illustration of the power that climate change can deliver – today – to our doorsteps.

The science is clear: climate change is here, now.

Ask the homeowners along the New Jersey and New York shores still homeless. Ask the local governments struggling weeks later to turn on power to their cold, darkened towns and cities. Ask the entire North East Coast, reeling from a catastrophe whose cost is estimated at $50 billion and rising. (I am not brave enough to ask those who've lost husbands or wives, children or grandparents).

I bring up these facts sadly, as one who has urged us to heed the scientific evidence on climate change for the past 25 years. The science is clear: climate change is here, now.

Superstorm Sandy is not the first storm, and certainly won't be the last. Still, it is hard for us as individual human beings to connect the dots. That's where observation, data and scientific analysis help us see.

No credible scientist disputes that we have warmed our climate by almost 1.5 C over land areas in the past century, most of that in the past 30 years.

As my colleagues and I demonstrated in a peer-reviewed study published this summer, climate extremes are already occurring much more frequently in the world we have warmed through our reliance on fossil fuels.

Our analysis showed that extreme summer heat anomalies used to be infrequent: covering only 0.1-0.2% of the globe in any given summer during the base period of our study, from 1951 to 1980. During the past decade, as the average global temperature rose, such extremes have covered 10% of the land.

Extreme temperatures deliver more than heat

The water cycle is especially sensitive to rising temperatures. Increased heat speeds up evaporation, causing more extreme droughts, like the $5 billion (and counting) drought in Texas and Oklahoma. It is linked to an expanding wildfire season and an increase by several fold in the frequency of large fires in the American west.

The heat also leads to more extreme sea surface temperatures – a key culprit behind Sandy's devastating force. The latent heat in atmospheric water vapour is the fuel that powers tornadoes, thunderstorms, and hurricanes. Stepping up evaporation with warmer temperatures is like stepping on the gas: More energy-rich vapour condenses into water drops, releasing more latent heat as it does so, causing more powerful storms, increased rainfall and more extreme flooding. This is not a matter of belief. This is high-school science class.

> *We must make the price of fossil fuels honest, reflecting their cost to society including the economic devastation wrought by storms like Hurricane Sandy....*
> *We can fix this. The answer is a price on carbon.*

The chances of getting a late October hurricane in New York without the help of global warming are extremely small. In that sense, you can blame Sandy on global warming. Sandy was the strongest recorded storm, measured by barometric pressure, to make landfall north of Cape Hatteras, eclipsing the hurricane of 1938.

But this fixation on determining the blame for a particular storm, or disputing the causal link between climate change and this or that storm, is misguided.

A better path forward means listening to the growing chorus – Sandy, extreme droughts and wildfires, intense rainstorms, record-breaking melting of Arctic sea ice – and taking action. Think of it like taking out an insurance policy for the planet.

We can fix this. The answer is a price on carbon. We must make the price of fossil fuels honest, reflecting their cost to society including the economic devastation wrought by storms like Sandy, the toll on farmland and ecosystems, as well as priceless human lives.

Whether that price takes the shape of a carbon tax, as some in Washington are now willing to discuss, or a carbon fee, as I have advocated, a price on carbon lets the market find the most effective ways to phase out our reliance on fossil fuels. It also moves us to a sustainable energy future where energy choices are made by individuals and communities, not by Washington mandates and lobbyists.

A price on carbon lets the market find the most effective ways to phase out... fossil fuels. It also moves us to a sustainable energy future where energy choices are made by individuals and communities.

A carbon fee, collected from fossil fuel companies, will increase consumer costs. So the money that is collected should be distributed to the public. As people try to minimise their energy costs to keep money for other things, their actions will stimulate the economy, drive innovations and transition us away from fossil fuels.

If we make our demand for action clear enough, I am optimistic that our leaders in Washington can look beyond the short-term challenges of today to see the looming, long-term threats ahead, and the answer that is right in front of them. We can't simply allow the next news cycle to distract us from the real task ahead.

Back in the 1980s, I introduced the concept of "climate dice" to make clear the difference between natural variability and climate-change driven extremes. As I predicted, the climate dice in the twenty-first century are now "loaded." It's not just bad luck that Sandy pummeled America's coasts, extreme drought devastated its midlands and wildfires scorched its mountains.

We loaded the dice. We changed our climate.

Dr. James E Hansen is a world acclaimed authority on global climate. He directs NASA's Goddard Institute for Space Studies.

Latest IPCC report finds, "Climate Change Now Felt on all Continents and Across the Oceans"

Changes in climate have already caused impacts on natural and human systems

by Suzanne Goldenberg Yokohama, Japan
The Guardian (London)
28 March 2014

Smoke billowing from a plant in Tokyo Bay, Japan. Government officials and scientists are gathered in Yokohama this week ahead of the launch of the IPCC report.

Photograph: Franck Robichon/EPA

CLIMATE CHANGE HAS ALREADY LEFT ITS MARK "ON ALL CONTINENTS and across the oceans," damaging food crops, spreading disease, and melting glaciers, according to the leaked text of a blockbuster UN climate science report due out on Monday.

Government officials and scientists are gathered in Yokohama this week to wrangle over every line of a summary of the report before the final wording is released on Monday – the first update in seven years.

Nearly 500 people must sign off on the exact wording of the summary, including the 66 expert authors, 271 officials from 115 countries, and 57 observers.

But governments have already signed off on the critical finding that climate change is already having an effect, and that even a small amount of warming in the future could lead to "abrupt and irreversible changes", according to documents seen by The Guardian (London).

"In recent decades, changes in climate have caused impacts on natural and human systems on all continents and across the oceans," the final report from the Intergovernmental Panel on Climate Change will say.

Climate change has already left its mark
"on all continents and across the oceans,"
damaging food crops, spreading disease,
and melting glaciers.

Some parts of the world could soon be at a tipping point. For others, that tipping point has already arrived. "Both warm water coral reef and Arctic ecosystems are already experiencing irreversible regime shifts," the approved version of the report will say.

This will be the second of three reports on the causes, consequences of and solutions to climate change, drawing on researchers from around the world.

The first report, released last September in Stockholm, found humans were the "dominant cause" of climate change, and warned that much of the world's fossil fuel reserves would have to stay in the ground to avoid catastrophic climate change.

This report will, for the first time, look at the effects of climate change as a series of risks – with those risks multiplying as temperatures warm.

The thinking behind the decision was to encourage governments to prepare for the full range of potential consequences under climate change.

Japanese environment minister, Nobuteru Ishihara, (third from the left) delivers a speech at the opening session of the IPCC working group II in Yokohama.

Photograph: Yoshikazu Tsuno/AFP/Getty Images

"It's much more about what are the smart things to do, then what do we know with absolute certainty," said Chris Field, one of the co-chairs overseeing the report.

"If we want to take a smart approach to the future, we need to consider a full range of possible outcomes and that means not only the more likely outcomes, but also outcomes for truly catastrophic impacts, even if those are lower probability," he said.

The gravest of those risks was to people in low-lying coastal areas and on small islands, because of storm surges, coastal flooding and sea-level rise.

The new IPCC report declares, "People living in large urban areas would be at risk from inland flooding that wipes out homes and businesses, water treatment centers and power plants, as well as from extreme heat waves.

"Food production was also at risk... from drought, flooding, and changing rainfall patterns. Crop yields could decline by 2% a decade over the rest of the century.

"Fisheries will be affected, with ocean chemistry thrown off balance by climate change. Some fish in the tropics could become extinct. Other species, especially in northern latitudes, are on the move.

"Drought could put safe drinking water in short supply. Storms could wipe out electricity stations, and damage other infrastructure...."

But people living in large urban areas would also be at risk from inland flooding that wipes out homes and businesses, water treatment centers and power plants, as well as from extreme heat waves.

Food production was also at risk, the report said, from drought, flooding, and changing rainfall patterns. Crop yields could decline by 2% a decade over the rest of the century.

Fisheries will also be affected, with ocean chemistry thrown off balance by climate change. Some fish in the tropics could become extinct. Other species, especially in northern latitudes, are on the move.

Drought could put safe drinking water in short supply. Storms could wipe out electricity stations, and damage other infrastructure, the report is expected to say.

Those risks will not be borne equally, according to draft versions of the report circulated before the meeting. The poor, the young and the elderly in all countries will all be more vulnerable to climate risks.

Climate change will slow down economic growth, and create new "poverty traps." Some areas of the world will also be more vulnerable – such as south Asia and south-east Asia.

The biggest potential risk, however, was of a number of those scenarios unfolding at the same time, leading to conflicts and wars, or turning regional problem into a global crisis, said Saleemul Haq, a senior fellow of the International Institute for Environment and Development and one of the authors of the report.

"The really scary impacts are when things start getting together globally," he said. "If you have a crisis in two or three places around the world, suddenly it's not a local crisis. It is a global crisis, and the repercussions of things going bad in several different places are very severe."

The report argues that the likelihood and potential consequences of many of these risks could be lowered if ambitious action is taken to reduce the greenhouse gas emissions that cause climate change. It also finds that governments – if they act now – can help protect populations from those risks.

But the report also acknowledges that a certain amount of warming is already locked in, and that in some instances there is no way to escape the effects of climate change.

The 2007 report on the effects of climate change contained an error that damaged the credibility of the UN climate panel, the erroneous claim that Himalayan glaciers could melt by 2035.

This year's report will be subject to far more rigorous scrutiny, scientists said. It will also benefit from an explosion of scientific research. The number of scientific publications on the impacts of climate change doubled between 2005 and 2010, the report will say.

Researchers said they also hoped to bring a fresh take on the issue. They said they hoped the reframing of the issue as a series of risks would help governments respond more rapidly to climate change.

> *The report also says, "Climate change will slow down economic growth and create new 'poverty traps.' Some areas of the world will also be more vulnerable – such as south Asia and south-east Asia.*
>
> *"The biggest potential risk, however, was of a number of those scenarios unfolding at the same time, leading to conflicts and wars, or turning regional problem into a global crisis...."*

"Previously the IPCC was accused of being very conservative," said Gary Yohe, professor of economics and environmental studies at Wesleyan University, one of the authors of the report. "This allows them to be less conservative without being open to criticism that they are just trying to scare people to death."

Sweeping Intelligence Report Sounds Alarm on Climate Threats

By Ben Geman
The Hill, Washington, DC
December 10, 2012

A NEW U.S. INTELLIGENCE COMMUNITY REPORT finds that climate change will fuel new conflicts and competition for resources in coming decades.

The federal National Intelligence Council's "Global Trends 2030" report issued Monday follows Defense Department analyses that have similarly called climate change an emerging security risk.

"Demand for food, water, and energy will grow by approximately 35, 40, and 50 percent respectively owing to an increase in the global population and the consumption patterns of an expanding middle class. Climate change will worsen the outlook for the availability of these critical resources," the report states.

> *"Demand for food, water, and energy will grow by approximately 35, 40, and 50 percent respectively owing to an increase in the global population and the consumption patterns of an expanding middle class. Climate change will worsen the outlook for the availability of these critical resources," the report states.*

Elsewhere, it notes that climate change is among the factors that will drive conflict in some regions.

"[M]any developing and fragile states — such as in Sub-Saharan Africa — face increasing strains from resource constraints and climate change, pitting different tribal and ethnic groups against one another and accentuating the separation of various identities," the report states.

Climate change will also drive migration patterns, analysts say.

From the report: "Internal migration — which will be at even higher levels than international migration — will be driven by rapid urbanization in the developing world and, in some countries toward the end of our time frame, by environmental factors and the impact of climate change."

Climate-change-driven migration is likely to affect Africa and Asia far more than other continents because of dependence on agriculture in Africa and parts of Asia and because of greater susceptibility in Asia to extreme weather events.

Dramatic and unforeseen changes already are occurring at a faster rate than expected. Most scientists are not confident of being able to predict such events. Rapid changes in precipitation patterns — such as monsoons in India and the rest of Asia — could sharply disrupt that region's ability to feed its population....

It also lists the possibility of much more rapid climate change among eight potential "black swan" events that could have the greatest "disruptive" effect.

"Dramatic and unforeseen changes already are occurring at a faster rate than expected. Most scientists are not confident of being able to predict such events. Rapid changes in precipitation patterns — such as monsoons in India and the rest of Asia — could sharply disrupt that region's ability to feed its population," the report states.

The overall report, the fifth such study issued since 1996, is a sweeping look at potential geopolitical changes and what's driving them.

It projects that the U.S. will remain the world's biggest power but that no nation will dominate as the "unipolar moment" ends.

"The US most likely will remain 'first among equals' among the other great powers in 2030 because of its preeminence across a range of power dimensions and legacies of its leadership role," it states.

"By 2030, no country — whether the US, China, or any other large country — will "be a hegemonic power," the report finds, adding:

> "[W]ith the rapid rise of multiple other powers, the 'unipolar moment' is over and *Pax Americana* – the era of unrivalled American ascendancy in international politics that began in 1945 – is fast winding down."

The World Bank

World Bank's Climate Change Report Says,
"Turn Down the Heat on Warming Planet"

By Anna Yukhananov
Reuters News Service
November 18, 2012

WASHINGTON, Nov 18 (Reuters) - "ALL NATIONS WILL SUFFER THE EFFECTS OF A WARMER world, but it is the world's poorest countries that will be hit hardest by food shortages, rising sea levels, cyclones and drought," the World Bank said in a report on climate change.

Under new World Bank President Jim Yong Kim, the global development lender has launched a more aggressive stance to integrate climate change into development.

"We will never end poverty if we don't tackle climate change. It is one of the single biggest challenges to social justice today," Kim told reporters on a conference call on Friday.

We will never end poverty if we don't tackle climate change. It is one of the single biggest challenges to social justice today,

The report, called "Turn Down the Heat," highlights the devastating impact of a world hotter by 4 degrees Celsius (7.2 Fahrenheit) by the end of the century, a likely scenario under current policies, according to the report.

Climate change is already having an effect: Arctic sea ice reached a record minimum in September, and extreme heat waves and drought in the last decade have hit places like the United States and Russia more often than would be expected from historical records, the report said.

Such extreme weather is likely to become the "new normal" if the temperature rises by 4 degrees, according to the World Bank report. This is likely to happen if not all countries comply with pledges they have made to reduce greenhouse gas emissions. Even assuming full compliance, the world will warm by more than 3 degrees by 2100.

Climate change is already having an effect: Arctic sea ice reached a record minimum in September, and extreme heat waves and drought in the last decade have hit places like the United States and Russia more often than would be expected from historical records.

In this hotter climate, the level of the sea would rise by up to 3 feet, flooding cities in places like Vietnam and Bangladesh. Water scarcity and falling crop yields would exacerbate hunger and poverty. Extreme heat waves would devastate broad swaths of the earth's land, from the Middle East to the United States, the report says. The warmest July in the Mediterranean could be 9 degrees hotter than it is today – akin to temperatures seen in the Libyan desert.

"It is my hope that this report shocks us into action," Kim, writes in the report. Scientists are convinced that global warming in the past century is caused by increasing concentrations of greenhouse gases produced by human activities such as the burning of fossil fuels and deforestation. These findings by the UN's Intergovernmental Panel on Climate Change were recognized by the national science academies of all major industrialized nations in a joint statement in 2010.

The combined effect of all these changes could be even worse, with unpredictable effects that people may not be able to adapt to, said John Schellenhuber, director of the Potsdam Institute for Climate Impact Research, which along with Climate Analytics prepared the report for the World Bank.

"If you look at all these things together, like organs cooperating in a human body, you can think about acceleration of this dilemma," said Schellenhuber, who studied chaos theory as a physicist. "The picture reads that this is not where we want the world to go."

Scientists are convinced that global warming... is caused by increasing concentrations of greenhouse gases produced by human activities such as the burning of fossil fuels and deforestation. These findings... were recognized by the national science academies of all major industrialized nations in a joint statement in 2010.

SHOCKED INTO ACTION

As the first scientist to head the World Bank, Kim has pointed to "unequivocal" scientific evidence for man-made climate change to urge countries to do more.

Kim said 97 percent of scientists agree on the reality of climate change.

Kim said the World Bank plans to further meld climate change with development in its programs.

Last year, the Bank doubled its funding for countries seeking to adapt to climate change, and now operates $7.2 billion in climate investment funds in 48 countries.

The World Bank study comes as almost 200 nations will meet in Doha, Qatar, from Nov. 26 to Dec. 7 to try to extend the Kyoto Protocol, the existing plan for curbing greenhouse gas emissions by developed nations that runs to the end of the year.

They have been trying off and on since Kyoto was agreed in 1997 to widen limits on emissions but have been unable to find a formula acceptable to both rich and poor nations.

Emerging countries like China, the world's biggest emitter of greenhouse gases, have said the main responsibility to cut emissions lies with developed nations, which had a headstart in sparking global warming.

Combating climate change also poses a challenge for the poverty-fighting World Bank: how to balance global warming with immediate energy needs in poor countries.

There really is no alternative to urgent action given the devastating consequences of climate change.... Now the question for the World Bank is how it will ensure that all of its investments respond to the imperatives of the report.

In 2010, the World Bank approved a $3.75 billion loan to develop a coal-fired power plant in South Africa despite lack of support from the United States, Netherlands and Britain due to environmental concerns.

"There really is no alternative to urgent action given the devastating consequences of climate change," global development group Oxfam said in a statement. "Now the question for the World Bank is how it will ensure that all of its investments respond to the imperatives of the report."

Kim said the World Bank now tries to avoid investing in coal unless there are no other options.

"But at the same time, we are the group of last resort in finding needed energy in countries that are desperately in search of it," he said.

The Center for Naval Analysis

National Security and the Accelerating Risks of Climate Change

The Center for Naval Analysis' (CNA) Military Advisory Board (MAB) is a distinguished group of retired senior flag and general officers from the Army, Navy, Air Force, Marine Corps, and Coast Guard who study the nexus between national security, climate change, and energy security, and work to inform and educate the public and policy makers on those issues.

For nearly a decade more than 30 generals and admirals have served on the board, using their knowledge and insight as military leaders to assess future risks to national security and explore options for mitigating those risks. Their efforts have led to five landmark reports, including _National Security and the Accelerating Risks of Climate Change_, released in the spring of 2014.

Foreward to the Official Report

(abridged for space)

May, 2014

Projected climate change is a complex challenge. Without action to build resilience, it will increase security risks over much of the planet. It will not only increase threats to developing nations... but it will also test the security of nations with robust capability, including significant elements of our National Power here at home....

When it comes to thinking through long-term global challenges, none are more qualified than our most senior military leaders.... It is through this analytical prism that 11 retired Generals and Admirals came together... to examine the security implications of climate change.

The Military Advisory Board has gathered to re-examine... climate change and national security. This update reflects their decades of experience as risk managers and geopolitical security experts.... The report deserves strong attention from not only the security community, but also from the government and the American public.

The update serves as a bipartisan call to action. It makes a compelling case that climate change is no longer a future threat—it is taking place now. ...

The update makes clear that actions... against the impacts of climate change are required today. We no longer have the option to wait and see. We applaud this group of American patriots for this important update. We commend its reading in full and its recommendations to the Administration, to Congress, and to the American people.

Signed,

Michael Chertoff
Former Secretary, Homeland Security

Leon Panetta
Former Secretary, Department of Defense

Executive Summary

See the Full Report at The Center for Naval Analysis

CNA's Military Advisory Board (MAB) first addressed the national security implications of climate change in our 2007 report — *National Security and the Threat of Climate Change.* We gather again as a group of 16 retired Generals and Admirals from the Army, Navy, Air Force, and Marine Corps to re-examine climate change in the context of a more informed, but more complex and integrated world, and to provide an update to our 2007 findings.

We are compelled to conduct this update now because of nearly seven years of developments in scientific climate projections; observed climate changes, particularly in the Arctic; the toll of observed extreme weather events both at home and abroad; and changes in the global security environment. Although we have seen some movement in mitigation and other areas where climate adaptation and resilience are starting to be included in planning documents, we gather again because of our growing concern over the lack of comprehensive action by both the United States and the international community to address the full spectrum of projected climate change issues.

The specific questions addressed in this update are:

1. Have new threats or opportunities associated with projected climate change or its effects emerged since our last report? What will be the impacts on our military?

2. The 2014 National Climate Assessment indicates that climate change, once considered an issue for a distant future, has moved firmly into the present. What additional responses should the national security community take to reduce the risks posed to our nation and to the elements of our National Power (Political, Military, Social, Infrastructure, and Information systems?

MAJOR FINDINGS:

Actions by the United States and the international community have been insufficient to adapt to the challenges associated with projected climate change. Strengthening resilience to climate impacts already locked into the system is critical, but this will reduce longterm risk only if improvements in resilience are accompanied by actionable agreements on ways to stabilize climate change.

Scientists around the globe are increasing their confidence, narrowing their projections, and reaffirming the likely causes of climate change.... Heat-trapping gases already in the atmosphere have committed us to a hotter future with more climate-related impacts over the next few decades. The magnitude of climate change beyond the next few decades depends on the amount of heat-trapping gases emitted globally, now and in the future." Some in the political realm continue to debate the cause of a warming planet and demand more data. Yet MAB member General Gordon Sullivan, U.S. Army (Ret.), has noted: "Speaking as a soldier, we never have 100 percent certainty. If you wait until you have 100 percent certainty, something bad is going to happen on the battlefield."

Climate mitigation and adaptation efforts are emerging in various places around the world, but the extent of these efforts to mitigate and adapt... are insufficient to avoid significant potential water, food, and energy insecurity; political instability; extreme weather events; and other manifestations of climate change. Coordinated, wide-scale, and well-executed actions to limit heat-trapping gases and to help prevent and protect against the worst projected climate change impacts are required—now.

The security ramifications of global climate change should be serving as catalysts for cooperation and change. Instead, climate change impacts are already accelerating instability in vulnerable areas of the world and are serving as catalysts for conflict.

> "... the projected impacts of climate change will be more than threat multipliers; they will serve as catalysts for instability and conflict."

Rapid population growth, especially in coastal and urban areas, and complex changes in the global security environment have made understanding the strategic security risks of projected climate changes more challenging. When it comes to thinking about the impacts of climate change, we must guard against a failure of imagination.

> "Climate change impacts transcend international borders and geographic areas of responsibility."

Accelerated melting of "old ice" in the Arctic is making the region more accessible to a wide variety of human activities, including shipping, resource extraction, fisheries, tourism, and other commerce. This activity level will accelerate in the coming decades. The United States and the international community are not prepared for the pace of change in the Arctic.

As the world's population and living standards grow, the projected climate impacts on the nexus of water, food, and energy security become more profound. Fresh water, food, and energy are inextricably linked, and the choices made over how these finite resources will be produced, distributed, and used will have increasing security implications.

> "... stresses on the water-food-energy nexus are a mounting security concern across a growing segment of the world."

Projected climate change impacts inside the United States will challenge key elements of our National Power and encumber homeland security. Of particular concern are climate impacts to our military, infrastructure, economic, and social support systems.

> "... impacts of climate change will strain our military forces in the coming decades."

In a security context, National Power is the ability to remain sovereign, protect national assets, and influence the behavior of others toward a desired outcome. Although the United States has embraced a more complex construct of National Power, a series of formal policy documents have introduced contrasting models of power, indicating that National Power has multiple and overlapping sources. In one of its simplest paradigms, National Power is modeled in terms of the ability to exert pressure through diplomatic, informational, military, and economic means (DIME). We are concerned about how projected climate change could degrade our National Power/PMESII.

RECOMMENDATIONS:

1. To lower national security risks, the United States should take global leadership role in preparing for the projected impacts of climate change.

This leadership role includes working with other nations, as well as with emerging nongovernmental and intergovernmental stakeholders —such as the Group of Seven (G-7), the World Trade Organization (WTO), —to build resilience for projected impacts of climate change. At the same time, the U.S. should lead global efforts to develop sustainable and more efficient energy solutions to help slow climate change.

2. Supported by National Intelligence Estimates, the U.S. military's Commanders should factor in the impacts of projected climate change across their full spectrum of planning and operations.

With partner nations, CCDRs should focus on building capacity and sustained resilience. Across areas of responsibility, they should work with nations and intergovernmental stakeholders to lower risk where the impacts of climate change likely will serve as a catalyst for conflict.

3. The United States should accelerate and consolidate its efforts to prepare for increased access and military operations in the Arctic.

DOD and other U.S. agencies should build on and accelerate plans recently put forward in Arctic strategic planning.... The Arctic is already becoming viable for commercial shipping and increased resource exploit-ation. The time to act is now. To expedite crisis response, the Arctic region should be assigned to one CCMD. To provide the U.S. with better standing in future disputes in the Arctic, the U.S. should become a signatory to the U.N. Convention on the Law of the Sea.

4. Climate adaptation planning should consider the water-food-energy nexus to ensure comprehensive decision making.

Rapidly growing population and urbanization combined with changes in weather patterns, will stress resource production and distribution, particularly water, food, and energy. These resources are linked, and adaptation planning must consider their interrelationships.

5. The projected impacts of climate change should be integrated fully into the National Infrastructure Protection Plan and the Strategic National Risk Assessment.

As military leaders, we know that we cannot wait for certainty. The failure to include a range of probabilities because it is not precise is unacceptable. The Strategic National Risk Assessment must include projected impacts of climate change over coming decades so that resilience requirements associated with these projections can be better defined in the National Infrastructure Protection Plan.

6. In addition to DOD's conducting comprehensive assessments of the impacts of climate change on mission and operational resilience, the Department should develop, fund, and implement plans to adapt....

This recommendation includes decisions to be made through any future processes, including base realignment and closure (BRAC), as well as expanding climate projections in planning and design for new bases, or other infra-structure. In new or existing bases, DOD should explore innovative solutions such as public-private partnerships to build climate change–resilient infrastructure. Climate change impacts should be considered in all vulnerability assessments, now and going forward...

"Corporations have been the principal economic actors for a long time; now they are the principal political actors as well. Neither the environment nor society fares well under corporatocracy. Environmentalists need to embrace public financing of elections, lobbying regulation, nonpartisan Congressional redistricting and other reforms as a core of their agenda. Today's politics will never deliver environmental sustainability."

"Our best hope for change is a fusion of those concerned about environmental sustainability, social justice and political democracy into one progressive force."

Global Warming and Modern Capitalism

By James Gustave Speth
September 17, 2008

http://www.thenation.com/doc/20081006/speth/print?rel=nofollow

This article is adapted from James Gustave Speth's *The Bridge at the Edge of the World: Capitalism, the Environment, and Crossing From Crisis to Sustainability* (Yale Univ. Press, 2013).

I GREW UP IN A SMALL TOWN ON THE EDISTO RIVER IN SOUTH CAROLINA in the 1940s and '50s. As a boy, I often swam the Edisto, though at first I could not buck the river's current. But as I grew older and stronger, I was able to make good headway against it. In my environmental work for close to four decades, I've always assumed America's environmental community would do the same – get stronger and prevail against the current. But in the past few years I have come to the conclusion that this assumption is incorrect. The environmental community has grown in strength and sophistication, but the environment has continued to deteriorate. The current has strengthened faster than we have and become more treacherous. It is time to consider what to do besides swimming against it.

It is no accident that environmental crisis is gathering as social injustice is deepening and growing inequality is impairing democratic institutions. Each is the result of a system of political economy–today's capitalism–that is profoundly committed to profits and growth and profoundly indifferent to nature and society. Left uncorrected, it is an inherently ruthless, rapacious system, and it is up to citizens, acting mainly through government, to inject human and natural values into that system. But this effort fails because progressive politics are too feeble and Washington is more and more in the hands of powerful corporations and great wealth. The best hope for change in America is a fusion of those concerned about the environment, social justice and strong democracy into one powerful progressive force. This fusion must occur before it is too late.

Sadly, while environmentalists have been winning many battles, we are losing the planet. Half the world's tropical and temperate forests are gone. The rate of deforestation in the tropics is about an acre a second. Half the planet's wetlands are gone. An estimated 90 percent of the large predator fish are gone and 75 percent of marine fisheries are overfished, fished to capacity or depleted, up from 5 percent a few decades ago. Twenty percent of the corals are gone; another 20 percent severely threatened. Species are disappearing about 1,000 times faster than normal. The planet has not seen such a spasm of extinction in 65 million years, since the dinosaurs disappeared. Each year desertification claims a Nebraska-sized area of productive capacity worldwide. Toxic chemicals can be found by the dozens in essentially every one of us.

The best hope for change in America is a fusion of those concerned about the environment, social justice and strong democracy into one powerful progressive force. This fusion must occur before it is too late.

Earth's ozone layer was severely depleted before the change was discovered. Human activities have pushed atmospheric carbon dioxide levels up by more than a third and have started the most dangerous change of all--planetary warming and climate disruption. Earth's ice fields are melting. Industrial processes are fixing nitrogen, making it biologically active, at a rate equal to nature's; one result is the development of hundreds of dead zones in the oceans because of overfertilization. Withdrawals of fresh water consume more than half of accessible runoff, and water shortages are multiplying here and abroad. The following rivers no longer reach the oceans in the dry season: the Colorado, Yellow, Ganges and Nile, among many others.

The United States–responsible for about 30 percent of the carbon dioxide added to the atmosphere–is, of course, deeply complicit in these global trends, and four decades of environmental effort have not stemmed the tide of decline. The United States is losing 6,000 acres of open space every day, and 100,000 acres of wetlands every year. Forty percent of US fish species are threatened with extinction, a third of plants and amphibians, 15 to 20 percent of birds and mammals. Half of US lakes and a third of the rivers still fail to meet the standards that the 1972 Clean Water Act said should be met by 1983, and a third of Americans live in counties that fail to meet EPA air-quality standards. We have done little to curb our wasteful energy habits or our steady population growth.

All we have to do to destroy the planet's climate and biota and leave a ruined world to our children and grandchildren is to keep doing exactly what we are doing.... Just continue to release greenhouse gases at current rates, impoverish ecosystems and release toxic chemicals at current rates, and the world in the latter part of this century won't be fit to live in. But human activities are not holding at current levels– they are accelerating dramatically.

The world economy has more than quadrupled since 1960 and is projected to quadruple again by mid century. At recent rates of growth, it will double in fifteen to seventeen years. It took all of human history to grow the $7 trillion world economy of 1950. We now grow by that amount in a decade. Societies face the prospect of enormous environmental deterioration just when they need to be moving strongly in the opposite direction.

The escalating processes of climate disruption, biotic impoverishment and toxification– which continue despite decades of warnings and earnest effort–are a severe indictment of capitalism. Capitalism as it is constituted today produces an economy and politics that are highly destructive to the environment. An unquestioning commitment to economic growth at any cost, powerful corporations whose overriding objective is to grow by generating profits (including profits from avoiding the environmental costs they create, from amassing deep subsidies and benefits from government and from continued deployment of technologies designed with little regard for the environment), markets that fail to recognize environmental costs unless corrected by government, government that is subservient to corporate interests and the growth imperative, rampant consumerism spurred by sophisticated advertising and marketing, economic activity so large in scale that it alters the fundamental biophysical operations of the planet--all combine to deliver an ever growing world economy that is undermining the ability of the planet to sustain life.

The escalating processes of climate disruption, biotic impoverishment and toxification – which continue despite decades of warnings and earnest effort – are a severe indictment of capitalism.

Mainstream environmentalism has proved largely incapable of coping with these forces. It works within the system--raising public awareness, offering responsive policies, lobbying and litigating. America has run a forty-year experiment on whether this environmentalism can succeed, and the results are in. The full burden of managing accumulating environmental threats has fallen to the environmental community, both in and outside government. But that burden is too great. The system of modern capitalism will grow in size and complexity and will generate ever larger environmental consequences, outstripping efforts to cope with them. Indeed, the system will seek to undermine those efforts and constrain them within narrow limits. Working only within the system will, in the end, not succeed. Transformative change in the system itself is needed.

The fundamental questions thus are about transforming capitalism as we know it. Can it be done? If so, how? And if not, what then? The good news is that there are a variety of prescriptions to take the economy and the environment off a collision course and to transform economic activity into something benign and restorative. The most important of these prescriptions range far beyond the traditional environmental agenda.

Market failure can be corrected by government, perverse subsidies can be eliminated and environmentally honest prices can be forged. The laws, incentives and governance structures under which corporations operate can be transformed to move from shareholder primacy to stakeholder primacy. But even more vital is the need to challenge economic growth and the consumerism it depends on. This challenge is as relevant to addressing social problems as environmental ones.

The never-ending drive to grow the economy undermines families, jobs, communities, the environment, ... even national security–but we are told that, in the end, we will somehow be better off. America has not applied its growth dividend to meeting social and environmental needs. There is good evidence that increased incomes do not lead to greater satisfaction with life. In affluent countries we have what might be called uneconomic growth, to borrow Herman Daly's phrase, where, if one could total up all the costs of growth, they would outweigh the benefits.

The fundamental questions... are about transforming capitalism as we know it.

Overriding commitment to economic growth--mere GDP growth – is consuming environmental and social capital, both in short supply. Affluent countries must become postgrowth societies where jobs and work life, the environment, communities and the public sector are no longer sacrificed to push up GDP.

There are many steps to slow growth while improving social and environmental well-being, such as: shorter workweeks and longer vacations; greater labor protections, job security and benefits; restrictions on advertising; a new design for the twenty-first-century corporation; strong social and environmental provisions in trade agreements; rigorous environmental and consumer protection, including full-cost pricing; greater economic and social equality, with progressive taxation of the rich and greater income support for the poor; heavy spending on public services and environmental amenities; a huge investment in education, skills and new technology; and initiatives to address population growth at home and abroad.

Instead of merely pursuing GDP growth, we need policies that address social needs directly– that strengthen families and communities and address the breakdown of social connectedness and the erosion of social capital; that guarantee good, well-paying jobs; that provide for universal healthcare and alleviate the devastating effects of mental illness; that provide a good education for all; that ensure care and companionship for the chronically ill and incapacitated; that recognize responsibilities to the half of humanity who live in poverty. There are many things that need to grow, and policy should concentrate there. Such measures, wise in their own right, should be seen as environmental measures too: central parts of the alternative to the destructive path we are on.

Americans are struggling with the combined impacts of higher food and fuel prices, crumbling financial assets, tighter credit and layoffs. These problems are not the result of a slowdown in GDP growth, and they will not be cured by more growth. Each is the result of government failing to intervene in the marketplace – in financial markets, in housing markets, in labor markets and elsewhere. As with climate change, we are on the receiving end of misguided policies that have led to deep structural maladies.

High prices are a problem because people don't have the money and alternatives (e.g., truly fuel-efficient vehicles) are not readily available. In a gutsy article in July, TIME noted that $4 gas was curbing sprawl, reducing pollution and traffic deaths, increasing fuel efficiency, and stimulating public transport, bike sales and walking. Honest prices would be higher for many things, but that does not mean Exxon should pocket the difference or that equity issues should remain unaddressed.

Conventional wisdom on the clash of economy and environment is that we can have it both ways, thanks to new technology. We do indeed need a revolution in the technologies of energy, transportation, construction, agriculture and more. But the rate of technological change required to deal with environmental challenges in the face of rapid economic growth is extremely high and rarely achieved. If pollution is cut in half but output doubles, there is no net gain. Housing, appliances and transportation can become more energy-efficient, but the improvements will be overwhelmed if there are more cars, larger houses and new appliances – and there are. There's a limit to how fast and far new technology can take us.

Psychological studies show that materialism is toxic to happiness and that more income and more possessions do not lead to a lasting sense of well-being or satisfaction....

Parallel to transcending our growth fetish, we must move beyond our consumerism and hyperventilating lifestyles. In the modern environmental era, there has been too little focus on consumption. This is slowly changing, but most mainstream environmentalists have not wanted to suggest that the positions they advocate would require serious personal changes. This reluctance to challenge consumption has been a big mistake, given the mounting environmental and social costs of American "affluenza," extravagance and wastefulness.

The good news is that more and more people sense that there's a great misdirection of life's energy. In a survey 83 percent of Americans say society is not focused on the right priorities, 81 percent say America is too focused on shopping and spending, 88 percent say American society is too materialistic, 74 percent believe excessive materialism is causing harm to the environment. If these numbers are correct, there's a powerful base to build on.

Psychological studies show that materialism is toxic to happiness and that more income and more possessions do not lead to a lasting sense of well-being or satisfaction with life. What makes people happy are warm personal relationships and giving rather than getting. Many people are trying to fight back against consumerism and commercialization. They say, Confront consumption. Practice sufficiency. Create social environments where overconsumption is viewed as silly, wasteful, ostentatious. Create commercial-free zones. Buy local. Eat slow food. Simplify your life. Downshift.

These prescriptions for change in the fundamental arrangements of capitalism are difficult, to put it mildly. What circumstances might make deep change plausible? A mounting sense of imminent crisis, wise leadership, the articulation of a new American narrative or story, as Bill Moyers has urged--all these would help. Most of all, we need a new politics and new social movement powerful enough to drive change.

Environmentalists must join social progressives to address the crisis of inequality unraveling our social fabric and undermining democracy. It is a crisis of soaring executive pay, huge incomes and increasingly concentrated wealth for a small minority while poverty rates approach a thirty-year high, wages stagnate despite rising productivity, social mobility and opportunity decline, the number of people without health insurance soars, job insecurity increases, safety nets shrink and Americans have the longest working day of all the rich countries. In an America with such vast social insecurity, where half the families just get by, economic arguments, even misleading ones, trump environmental ones.

> *There has been too little focus on consumption.... Most environmentalists have not wanted to suggest that [their] positions... require personal changes. This reluctance... has been a big mistake, given the mounting costs of American "affluenza," extravagance and wastefulness.*

Environmentalists must also join those seeking to reform politics and strengthen democracy. America's gaping social and economic inequality poses a grave threat to democracy. We are seeing the emergence of a vicious circle: income disparities shift political access and influence to wealthy constituencies and large businesses, which further imperils the potential of the democratic process to act to correct the economic disparities. Corporations have been the principal economic actors for a long time; now they are the principal political actors as well. Neither environment nor society fares well under corporatocracy. Environmentalists need to embrace public financing of elections, lobbying regulation, nonpartisan Congressional redistricting and other reforms as a core of their agenda. Today's politics will never deliver environmental sustainability.

My point of departure was the momentous environmental challenge we face. But today's environmental reality is linked powerfully with other realities, including growing social inequality and neglect and the erosion of democratic governance and popular control. So my conclusion is that we as citizens must mobilize our spiritual and political resources for transformative change on all three fronts. Our best hope for change is a fusion of those concerned about environmental sustainability, social justice and political democracy into one progressive force.

One area where fusion is beginning is the conversation between environmental and social justice activists on solutions, including green-collar ones, to the climate change threat. That's encouraging, but it's a small part of what's needed. Mostly, everyone is still in his or her silo. A sustained dialogue is urgently needed among the three communities, to build a common agenda for action and a shared commitment to build a new social movement for change in America. We are all communities of a shared fate. We will rise or fall together.

James Gustave Speth, dean of the Yale School of Forestry and Environmental Studies, is the author of *The Bridge at the Edge of the World: Capitalism, the Environment, and Crossing From Crisis to Sustainability* (Yale University Press, 2013).

Faith Leaders Need to Find Their Voice on Climate Change

Religious institutions need to set their moral compass on one of the great humanitarian issues of our time

By Christiana Figueres
The Guardian (London)
May 7, 2014

Pope Francis I greets pilgrims gathered at Saint Peter's Square, Vatican city, Rome

Photograph: Vincenzo Pinto/AFP/Getty Images

SAVING THE EARTH AND ITS PEOPLES FROM DANGEROUS CLIMATE CHANGE is an economic, social and environmental issue – and a moral and ethical one too that goes to the core of many, if not all of the world's great faiths.

Unchecked, the rise in greenhouse gas emissions is likely to visit ever higher levels of suffering on the vulnerable, the marginalised and indeed people everywhere.

The Himalayan country of Nepal, which I just visited, is a case in point: here unstable lakes are forming from melting glaciers high in the mountains. Some have already burst their banks sending vertical tsunamis down valleys washing away power lines, homes and lives.

Many forward-looking cities, progressive companies and concerned citizens are urging their governments to ink a new climate agreement in 2015.

It is time for faith groups and religious institutions to find their voice and set their moral compass on one of the great humanitarian issues of our time.

Overcoming poverty, caring for the sick and the infirm, feeding the hungry and a whole range of other faith-based concerns will only get harder in a climate challenged world.

In supporting greater ambition by nations, religious and faith groups can assist in shaping a world that is less polluted and healthier, safer and more secure for every man, woman and child.

There are a myriad of ways in which churches and mosques to synagogues and temples can assist towards an ambitious climate agreement.

A world-wide campaign by universities and cities, aimed at divesting pension and endowment funds from fossil fuel shares, is also gaining ground.

Saving the earth and its peoples from dangerous climate change is an economic, social and environmental issue – and a moral and ethical one too that goes to the core of many if not all of the world's great faiths.

South African Anglican Archbishop Desmond Tutu recently called for an anti-apartheid style boycott and disinvestment campaign against the fossil fuel industry.

Some smaller churches are already moving including in Australia and New Zealand. In the US, 12 religious institutions have already divested from the fossil fuel industry.

In 2013, the United Church of Christ (UCC) became the first national faith communion to vote to divest from fossil fuel companies, with the support of its major investment institution, United Church Funds (UCF). UCF manages investment funds of over 1,000 churches, conferences, associations and other ministries, with more than half a billion dollars in assets.

In February this year, the Trinity-St Paul's United Church in Toronto voted unanimously to ensure that its own funds are not invested in any of the world's 200 largest fossil fuel companies. Multi-faith groups in Australia and North America recently sent a letter to Pope Francis saying it is "immoral" to profit from fossil fuels.

The Church of England recently voted to review its investment policy in respect to fossil fuels – a step in the right direction and a potentially powerful signal to its 28 million followers. Divestment may be a question of morality, but it is prudent too. Experts estimate that green-house gas emissions will peak in around ten years' time and then come down sharply afterwards. ...

The good news is that governments have agreed to secure a new agreement on climate change when they meet in Paris, France, next year. If the world and its people are to be spared dangerous climate change that agreement needs to also be meaningful with polices for carbon neutrality in the second half of the century if a global temperature rise is to be kept under 2C.

Leaders of faith groups, from Christians and Muslims to Hindus, Jews and Buddhists have a responsibility and an opportunity over the next 18 months to provide a moral compass to their followers and to political, corporate, financial and local authority leaders.

In doing so, faiths and religions can not only secure a healthy and habitable world for all but contribute to the spiritual and physical well-being of humanity now and for generations to come.

NATIONAL ACADEMY OF SCIENCES
NATIONAL ACADEMY OF ENGINEERING
INSTITUTE OF MEDICINE
NATIONAL RESEARCH COUNCIL

FROM THE NATIONAL ACADEMIES

Date: Feb. 27, 2014   FOR IMMEDIATE RELEASE

U.S. National Academy of Sciences, U.K. Royal Society Issue Joint Publication on Climate Change

WASHINGTON -- The U.S. National Academy of Sciences and the Royal Society, the national science academy of the U.K., released a joint publication today in Washington, D.C., that explains the clear evidence that humans are causing the climate to change, and that addresses a variety of other key questions commonly asked about climate change science.

"As two of the world's leading scientific bodies, we feel a responsibility to evaluate and explain what is known about climate change, at least the physical side of it, to concerned citizens, educators, decision makers and leaders, and to advance public dialogue about how to respond to the threats of climate change," said NAS President Ralph J. Cicerone.

"Our aim with this new resource is to provide people with easy access to the latest scientific evidence on climate change, including where scientists agree and where uncertainty still remains," added Sir Paul Nurse, president of the Royal Society. "We have enough evidence to warrant action being taken on climate change; it is now time for the public debate to move forward to discuss what we can do to limit the impact on our lives and those of future generations."

Climate Change: Evidence and Causes, written and reviewed by leading experts in both countries, lays out which aspects of climate change are well-understood, and where there is still uncertainty and a need for more research.

> *If emissions continue unabated, future climate changes will substantially exceed those that have occurred so far...*

Carbon dioxide (CO^2) has risen to levels not seen for at least 800,000 years, and observational records dating back to the mid-19th century show a clear, long-term warming trend. The publication explains that measurements that distinguish between the different forms of carbon in the atmosphere provide clear evidence that the increased amount of CO^2 comes primarily from the combustion of fossil fuels, and discusses why the warming that has occurred along with the increase in CO^2 cannot be explained by natural causes such as variations in the sun's output.

The publication delves into other commonly asked questions about climate change, for example, what the slower rate of warming since the very warm year in 1998 means, and whether and how climate change affects the strength and frequency of extreme weather events.

Many effects of climate change have already become apparent in the observational record, but the possible extent of future impacts needs to be better understood. For example, while average global sea levels have risen about 8 inches (20 cm) since 1901, and are expected to continue to rise, more research is needed to more accurately predict the size of future sea-level rise. In addition, the chemical balance of the oceans has shifted toward a more acidic state, which makes it difficult for organisms such as corals and shellfish to form and maintain their shells. As the oceans continue to absorb CO^2, their acidity will continue to increase over the next century, along with as yet undetermined impacts on marine ecosystems and the food web.

Even if greenhouse gas emissions were to suddenly stop, it would take thousands of years for atmospheric CO^2 to return to levels before the industrial era. If emissions continue unabated, future climate changes will substantially exceed those that have occurred so far, the publication says.

The chemical balance of the oceans has shifted toward a more acidic state, which makes it difficult for organisms such as corals and shellfish to form and maintain their shells. As the oceans continue to absorb CO^2, their acidity will continue to increase over the next century, along with as yet undetermined impacts on marine ecosystems and the food web.

The authoring committee offers this brief explanation of the science of climate change to help inform policy debates about the choices available to nations and the global community for reducing the magnitude of climate change and adapting to its impacts.

*   *   *

The publication is available to download for free at www.nap.edu and as an interactive website can be found at http://nas-sites.org/americasclimatechoices/events/a-discussion-on-climate-change-evidence-and-causes.

The National Academy of Sciences (NAS) is a private, independent nonprofit institution that provides science, technology, and health policy advice under a congressional charter granted to NAS in 1863. For more information, visit http://national-academies.org.

The Royal Society is a self-governing fellowship of many of the world's most distinguished scientists drawn from all areas of science, engineering, and medicine. The society's fundamental purpose, reflected in its founding charters of the 1660s, is to recognize, promote, and support excellence in science and to encourage the development and use of science for the benefit of humanity. For further information, visit http://royalsociety.org.

Contacts:

William Kearney, Director of Media Relations
National Academy of Sciences
e-mail: news@nas.edu

Chloe McIvor, Press Officer
The Royal Society
e-mail: chloe.mcivor@royalsociety.org

Now Is the Time to Act on Climate Change

U.N. Secretary General Ban Ki Moon
New York, New York
September 2, 2014

CLIMATE CHANGE HAS BEEN ONE OF MY TOP PRIORITIES since the day I took office in 2007. I said then that if we care about our legacy for succeeding generations, this is the time for decisive global action. I have been pleased to see climate change rise on the political agenda and in the consciousness of people worldwide. But I remain alarmed that governments and businesses have still failed to act at the pace and scale needed.

Time is running out. The more we delay, the more we will pay. Climate change is accelerating and human activities are the principal cause, as documented in a series of authoritative scientific reports from the Intergovernmental Panel on Climate Change. The effects are already widespread, costly and consequential – to agriculture, water resources, human health, and ecosystems on land and in the oceans. Climate change poses sweeping risks for economic stability and the security of nations.

"Time is running out. The more we delay, the more we will pay."

I have traveled the world to see the impacts for myself, from the Arctic to the Antarctic, from the low-lying islands of the Pacific threatened by rising seas to the retreating glaciers of Greenland, the Andes and the Alps. I have seen encroaching deserts in Mongolia and the Sahel and endangered rain forests in Brazil. Everywhere I have talked with people on the front lines who are deeply concerned about the threat of climate change to their way of life and their future.

My travels have also introduced me to growing numbers of people – from heads of Government to business leaders – who are prepared to invest political and financial capital in the solutions we need. They understand that climate change is an issue for all people, all businesses, all governments. They recognize that we can avert the risks if we take decisive action now.

> *Climate change is not just an issue for*
> *the future, it is an urgent issue for today.*

Later this month, on September 23, I am convening a Climate Summit at the United Nations in New York. The Summit has two goals: to mobilize political will for a meaningful universal agreement at the climate negotiations in Paris in 2015; and to catalyze ambitious action on the ground to reduce greenhouse gas emissions and strengthen resilience to the changes that are already happening.

I have invited leaders from government, business, finance and civil society to present their vision, make bold announcements and forge new partnerships that will support the transformative change the world needs. The Summit will highlight a number of areas where we feel we can achieve the highest impact, as showcased in this pre-Summit series of blog posts by some of the most influential thinkers and actors in the climate arena.

Climate change is not just an issue for the future, it is an urgent issue for today. Instead of asking if we can afford to act, we should be asking what is stopping us, who is stopping us, and why?

Let us join forces to push back against skeptics and entrenched interests. Let us support the scientists, economists, entrepreneurs and investors who can persuade government leaders and policy-makers that now is the time for action.

This presentation is preliminary and preparatory for the UN Climate Summit 2014 to be held at the United Nation's headquarters in New York on September 23, 2014. Leaders from around the world will attend, including President Barack Obama, President Xi Jin-ping, from the People's Republic of China, and many other world leaders.

World Meteorological Organization
Weather • Climate • Water

Record Greenhouse Gas Levels Impact Atmosphere and Oceans

Press Release No. 1002
September 9, 2014

For use of the information media

Geneva, Switzerland | September 9, 2014 – The amount of greenhouse gases in the atmosphere reached a new record high in 2013, propelled by a surge in carbon dioxide levels. This is according to the World Meteorological Organization's annual Greenhouse Gas Bulletin, which injected greater urgency into the need for concerted international action against accelerating and potentially devastating climate change.

The "Greenhouse Gas Bulletin" showed that between 1990 and 2013 there was a 34% increase in radiative forcing – the warming effect on our climate – because of long-lived greenhouse gases such as carbon dioxide (CO_2), methane and nitrous oxide.

In 2013, concentration of CO_2 in the atmosphere was 142% of the pre-industrial era (1750), and of methane and nitrous oxide 253% and 121% respectively.

The observations from WMO's Global Atmosphere Watch (GAW) network showed that CO_2 levels increased more between 2012 and 2013 than during any other year since 1984. Preliminary data indicated that this was possibly related to reduced CO_2 uptake by the earth's biosphere in addition to the steadily increasing CO_2 emissions.

The WMO Greenhouse Gas Bulletin reports on atmospheric concentrations – and not emissions - of greenhouse gases. Emissions represent what goes into the atmosphere. Concentrations represent what remains in the atmosphere after the complex system of interactions between the atmosphere, biosphere and oceans. About a quarter of the total emissions are taken up by the oceans and another quarter by the biosphere, reducing the amount of CO_2 in the atmosphere.

The ocean cushions the increase in CO_2 that would otherwise occur in the atmosphere, but with far-reaching impacts. The current rate of ocean acidification appears unprecedented at least over the last 300 million years, according to an analysis in the report.

"We know without any doubt that our climate is changing and our weather is becoming more extreme due to human activities such as the burning of fossil fuels," said WMO Secretary-General Michel Jarraud.

"The Greenhouse Gas Bulletin shows that, far from falling, the concentration of carbon dioxide in the atmosphere actually increased last year at the fastest rate for nearly 30 years. We must reverse this trend by cutting emissions of CO_2 and other greenhouse gases across the board," he said. "We are running out of time."

"Carbon dioxide remains in the atmosphere for many hundreds of years and in the ocean for even longer. Past, present and future CO2 emissions will have a cumulative impact on global warming and ocean acidification. The laws of physics are non-negotiable," said Mr Jarraud.

"The Greenhouse Gas Bulletin provides a scientific base for decision-making. We have the knowledge and the tools for action to keep temperature increases within 2 degrees C to give our planet a chance and to give our children and grandchildren a future. Pleading ignorance can no longer be an excuse for not acting," said Mr Jarraud.

"The inclusion of a section on ocean acidification in this Greenhouse Gas Bulletin is appropriate and needed. It is high time the ocean, as the primary driver of the planet's climate, becomes a central part of climate change discussions," said Wendy Watson-Wright, Executive Secretary of the Intergovernmental Oceanographic Commission of UNESCO.

"If global warming is not a strong enough reason to cut CO2 emissions, ocean acidification should be, since its effects are already being felt and will increase for many decades to come. I echo WMO Secretary General Jarraud's concern – we ARE running out of time," she said.

Atmospheric Concentrations

Carbon dioxide accounted for 80% of the 34% increase in radiative forcing by long-lived greenhouse gases from 1990 to 2013, according to the U.S. National Oceanic and Atmospheric Administration (NOAA) Annual Greenhouse Gas Index. On the global scale, the amount of CO2 in the atmosphere reached 396.0 parts per million in 2013. The increase of CO2 from 2012 to 2013 was 2.9 parts per million, which is the largest annual increase for the period 1984-2013.

Methane is the second most important long-lived greenhouse gas. Approximately 40% of methane is emitted into the atmosphere by natural sources (e.g., wetlands and termites), and about 60 % comes from human activities like fossil fuel exploitation, cattle breeding, rice agriculture, landfills and biomass burning. Atmospheric methane reached a new high of about 1,824 parts per billion (ppb) in 2013, due to increased emissions from anthropogenic sources.

Nitrous oxide (N2O) is emitted into the atmosphere from both natural (about 60%) and anthropogenic sources (approximately 40%), including oceans, soil, biomass burning, fertilizer use, and various industrial processes. Its atmospheric concentration in 2013 was about 325 parts per billion. Its impact on climate, over a 100-year period, is 298 times greater than equal emissions of carbon dioxide.

Ocean Acidification: The ocean currently absorbs one-fourth of anthropogenic CO2 emissions, reducing the increase in atmospheric CO2 that would otherwise occur because of fossil fuel combustion. Enhanced ocean CO2 uptake alters the marine carbonate system and leads to increasing acidity. The ocean's acidity increase is already measurable as oceans take up about 4 kilograms of CO2 per day per person. The current rate of ocean acidification appears unprecedented over the last 300 million years.... In the future, acidification will continue to accelerate at least until mid-century, based on projections from Earth system models.

The potential consequences of ocean acidification are complex. A major concern is the response of calcifying organisms, such as corals, mollusks, and some plankton, because their ability to build shell or skeletal material depends on the abundance of carbonate ion. For many organisms, calcification declines with increased acidification. Other impacts of acidification include reduced survival, and growth rates as well as changes in physiological functions and reduced biodiversity.

Ocean Acidification: The Other Carbon Dioxide Problem

Welcome to the NOAA Ocean Acidification Program!
Ocean acidification is emerging as an urgent environmental and economic issue
on our nation's east and west coasts and in many areas of the world.

OCEAN ACIDIFICATION IS OCCURRING BECAUSE THE WORLD'S OCEANS are absorbing increasing amounts of atmospheric carbon dioxide, leading to greater acidity. This is threatening the fundamental chemical balance of ocean and coastal waters from pole to pole. Since the beginning of the industrial revolution, the release of carbon dioxide (CO_2) from humankind's industrial and agricultural activities has increased the amount of CO_2 in the atmosphere. The ocean absorbs about a quarter of the CO_2 we release into the atmosphere every year, so as atmospheric CO_2 levels increase, so do the levels in the ocean. Initially, many scientists focused on the benefits of the ocean removing this greenhouse gas from the atmosphere. However, decades of ocean observations now show that there is also a downside — the CO_2 absorbed by the ocean is changing the chemistry of the seawater, a process called "ocean acidification."

Ocean Acidification Mission statement

To understand the changing chemistry of the oceans and the impacts of ocean acidification on marine ecosystems, oceanographers have been studying how CO_2 emissions affect the ocean system for more than three decades and continue to monitor ocean acidification in all the world's oceans from coral reef ecosystems to deep North Pacific waters. Our mission group collects several types of carbon measurements throughout the world's oceans. We participate in large-scale research cruises across ocean basins and along coastlines at regular intervals to study how ocean chemistry is changing through time.

We also make measurements of the partial pressure of CO_2 (pCO_2) in the surface water of the world's oceans using automated analytical systems on moorings and underway platforms. We are currently in the process of adding pH, oxygen, chlorophyll, and turbidity sensors to our existing moored and underway systems to more accurately and precisely study the changes associated with ocean acidification.

The Biological Impacts

Ocean acidification is expected to impact ocean species to varying degrees. Photosynthetic algae and seagrasses may benefit from higher CO_2 conditions in the ocean, as they require CO_2 to live just like plants on land. On the other hand, studies have shown that a more acidic environment has a dramatic effect on some calcifying species, including oysters, clams, sea urchins, shallow water corals, deep sea corals, and calcareous plankton. When shelled organisms are at risk, the entire food web may also be at risk. Today, more than a billion people worldwide rely on food from the ocean as their primary source of protein. Many jobs and economies in the U.S. and around the world depend on the fish and shellfish in our oceans.

Shellfish

In recent years, there have been near total failures of developing oysters in both aquaculture facilities and natural ecosystems on the West Coast. These larval oyster failures appear to be correlated with naturally occurring upwelling events that bring low pH waters undersaturated in aragonite as well as other water quality changes to nearshore environments. Lower pH values occur naturally on the West Coast during upwelling events, but a recent observations indicate that anthropogenic CO_2 is contributing to seasonal undersaturation. Low pH may be a factor in the current oyster reproductive failure; however, more research is needed to disentangle potential acidification effects from other risk factors, such as episodic

Research on the effects of higher ocean acidity levels on shellfish is taking place in many locations.

freshwater inflow, pathogen increases, or low dissolved oxygen. It is premature to conclude that acidification is responsible for the recent oyster failures, but acidification is a potential factor in the current crisis to this $100 million a year industry, prompting new collaborations and accelerated research on ocean acidification and potential biological impacts.

Coral

Many marine organisms that produce calcium carbonate shells or skeletons are negatively impacted by increasing CO_2 levels and decreasing pH in seawater. For example, increasing ocean acidification has been shown to significantly reduce the ability of reef-building corals to produce their skeletons. In a recent paper, coral biologists reported that ocean acidification could compromise the successful fertilization, larval settlement and survivorship of Elkhorn coral, an endangered species. These research results suggest that ocean acidification could severely impact the ability of coral reefs to recover from disturbance. Other research indicates that, by the end of this century, coral reefs may erode faster than they can be rebuilt. This could compromise the long-term viability of these ecosystems and perhaps impact the estimated one million species that depend on coral reef habitat.

Ocean Acidification: An Emerging Global Problem

Since the start of the industrial revolution, the ocean has silently absorbed roughly 30% of the carbon dioxide that people generate through industry and agriculture; heating and transportation. Now ocean chemistry of the seawater is rapidly changing in a process known as ocean acidification. These changes in seawater chemistry affect animal growth, survival and behavior, and they are depleting the ocean of calcium carbonate, a nutrient vital for shellfish to build shells. Marine organisms with calcium carbonate shells or skeletons – such as corals, oysters, clams, and mussels – can be affected by small changes in acidity. That's important, because shelled organisms are essential throughout the marine food chain.

Over the last decade, there has been much focus in the ocean science community on the potential impacts of ocean acidification. Since sustained efforts to monitor ocean acidification worldwide are only beginning, it is currently impossible to predict exactly how ocean acidification will cascade throughout the marine food chain and affect the structure of marine ecosystems.

Alaska Dispatch NEWS

A wake-up call in Alaska about ocean acidification and coastal communities

by Jeremy Mathis and Steve Colt | July 29, 2014

A NEW STUDY SHOWS, FOR THE FIRST time, that ocean acidification is driving changes in waters vital to Alaska's commercial fisheries and traditional subsistence way of life.

As one of our planet's most under-recognized challenges, ocean acidification is emerging because the sea is absorbing increasing amounts of carbon dioxide from the atmosphere. CO2 concentrations are now higher than at any time during the past 800,000 years, and the current rate of increase is likely unprecedented in history. Ocean acidification is literally causing a sea change, threatening the fundamental health of ocean and coastal waters from pole to pole. And, as the new study indicates, the implications for Alaska may be profound.

OPINION: As ocean acidification emerges, it will affect Alaskans already at risk and coastal economies the most. The time to plan is now.   Loren Holmes photo

> *Ocean acidification is literally causing a sea change, threatening the fundamental health of ocean and coastal waters from pole to pole.*

Led by National Oceanic and Atmospheric Administration (NOAA) and University of Alaska scientists, the study stands out not just because of findings about the intensity of ocean acidification on Alaska's marine life, but because it assessed potential risk to Alaskan communities. The study considered social impacts such as food security, subsistence, jobs and educational opportunities. Findings indicate that communities in Alaska's southeast and southwest regions, those that are among the most important contributors to commercial and subsistence fisheries, are also the communities most at risk. Highly reliant on shellfish and salmon and other finfish, these communities have relatively lower income and employment options. Current changes in ocean chemistry and rising CO2 levels mean their vulnerability will only grow in the coming decades.

While more monitoring is needed to understand ocean acidification's implications for marine life, we see the new study as a wake-up call because changing seawater chemistry can be destructive to the skeletons and shells of many species. Ocean acidification makes it more challenging to build and maintain shells and, in some instances, even causes them to dissolve. Harvested shellfish such as clams and scallops will likely suffer from ocean acidification, and studies show that both red king crab and Tanner crab species grow more slowly and die more often in high-CO_2 waters. Risk to finfish may be lower but ocean acidification may cause changes that reduce their populations.

Of concern are Alaska's vulnerable coastal economies and highly valued ways of life. Alaska's fishing industry supports more than 100,000 jobs and generates more than $5 billion in annual revenue. About 120,000 Alaskans rely on subsistence fisheries for most, if not all, of their dietary protein. Fishery-related tourism brings in more than $300 million annually. Moreover, this state's highly productive fisheries are critical to America's global balance of trade. Alaska's coastline, which is 50 percent greater than all of the rest of our nation's coastline combined, produces about half of the total U.S. commercial catch. As with any investment, we have to understand the risks and vulnerabilities associated with our fisheries.

Ocean acidification is a global phenomenon. But the effects on species, mostly negative so far, are local.

There is a shared urgency, in Alaska and globally, to understand more about these risks and what ocean acidification means for lives, livelihoods and communities and regional and national economies. Ocean acidification is a global phenomenon. But the effects on species, mostly negative so far, are local. The new study is one step toward fuller understanding. It is presented with the hope that its scientific insights will be considered in addressing emerging challenges to already at-risk Alaska communities.

Building resiliency in Alaska means continued, critical monitoring of near-shore regions and having informed local policies that, ultimately, are developed by communities themselves. For individual communities, losing a major fishery and essential subsistence can be a game-changer. As ocean acidification emerges, so can adaptive and mitigation strategies that recognize community values and educate the public about risks and elements to offset them. The challenge is to continue reaping the benefits while also safeguarding traditional and contemporary uses of Alaska's and the world's fragile, finite marine resources.

Source URL: http://www.adn.com/article/20140729/wake-call-alaska-about-ocean-acidification-and-coastal-communities

Jeremy T. Mathis, Ph.D., *is an oceanographer at NOAA's Pacific Marine Environmental Laboratory in Seattle.* **Steve Colt, Ph.D.,** *is a professor of economics at the University of Alaska Anchorage. This editorial is drawn from a study published July 29, 2014 in the scientific journal, "Progress in Oceanography."*

The National Science Foundation

Record California Drought Directly Linked To Climate Change

by Joe Romm
September 29, 2014

"The drought crippling California is by some measures the worst in the state's history." Photo: NOAA

A STANFORD STUDY FUNDED BY THE NATIONAL SCIENCE FOUNDATION (NSF) confirms a growing body of research that finds "The atmospheric conditions associated with the unprecedented drought in California are very likely linked to human-caused climate change."

The NSF news release, headlined, "Cause of California drought linked to climate change," explains:

> Climate scientist Noah Diffenbaugh of Stanford University and colleagues used a novel combination of computer simulations and statistical techniques to show that a persistent region of high atmospheric pressure over the Pacific Ocean–one that diverted storms away from California–was much more likely to form in the presence of modern greenhouse gas concentrations.

Unprecedented droughts often combine a reduction in precipitation with higher temperatures that increase evaporation, leaving soil parched. As the NSF notes in this case, "Combined with unusually warm temperatures and stagnant air conditions, the lack of precipitation has triggered a dangerous increase in wildfires and incidents of air pollution across the state."

[Rising levels of] carbon pollution mean we'll be seeing more and more dangerous record droughts.

California's 3-year drought has reached epic proportions. The L.A. Times reported last week, "Drought has 14 communities on the brink of waterlessness."

Here's the most recent Drought Monitor for the state:

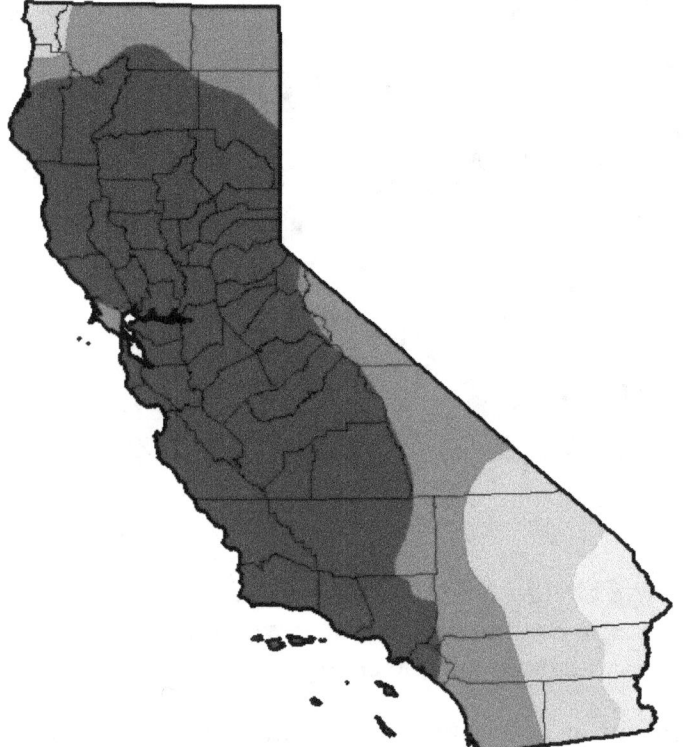

U.S. Drought Monitor
California

September 23, 2014
(Released Thursday, Sep. 25, 2014)
Valid 8 a.m. EDT

Drought Conditions (Percent Area)

	None	D0-D4	D1-D4	D2-D4	D3-D4	D4
Current	0.00	100.00	100.00	95.34	81.92	58.41
Last Week 9/16/2014	0.00	100.00	100.00	95.42	81.92	58.41
3 Months Ago 6/24/2014	0.00	100.00	100.00	100.00	76.69	32.98
Start of Calendar Year 12/31/2013	2.61	97.39	94.25	87.53	27.59	0.00
Start of Water Year 10/1/2013	2.63	97.37	95.95	84.12	11.36	0.00
One Year Ago 9/24/2013	2.63	97.37	96.04	89.84	11.36	0.00

Intensity:

D0 Abnormally Dry   D3 Extreme Drought

D1 Moderate Drought   D4 Exceptional Drought

D2 Severe Drought

The Drought Monitor focuses on broad-scale conditions. Local conditions may vary. See accompanying text summary for forecast statements.

Author:
Richard Heim
NCDC/NOAA

http://droughtmonitor.unl.edu/

The Impacts on Health from Climate Change

Weather and climate play a significant role in people's health. Changes in climate affect the average weather conditions that we are accustomed to. Warmer average temperatures will lead to hotter days and more frequent and longer heat waves. This could increase the number of heat-related illnesses and deaths. Increases in the frequency or severity of extreme weather events such as storms could increase the risk of dangerous flooding, high winds, and other direct threats to people and property. Warmer temperatures could increase the concentrations of unhealthy air and water pollutants. Changes in temperature, precipitation patterns, and extreme events could enhance the spread of some diseases.

The health impacts of climate change will depend on many factors. These factors include the effectiveness of a community's public health and safety systems to address or prepare for the risk and the behavior, age, gender, and economic status of individuals affected. Impacts will likely vary by region, the sensitivity of populations, the extent and length of exposure to climate change impacts, and society's ability to adapt to change.

Although the United States has well-developed public health systems (compared with those of many developing countries), climate change will still likely affect many Americans. In addition, the impacts of climate change on public health around the globe could have important consequences for the United States. For example, more frequent and intense storms may require more disaster relief and declines in agriculture may increase food shortages.

Key Points

* A warmer climate is expected to both increase the risk of heat-related illnesses and death and worsen conditions for air quality.

* Climate change will likely increase the frequency and strength of extreme events (such as floods, droughts, and storms) that threaten human safety and health.

* Climate changes may allow some diseases to spread more easily.

Impacts from Heat Waves

Heat waves can lead to heat stroke and dehydration, and are the most common cause of weather-related deaths. Excessive heat is more likely to impact populations in northern latitudes where people are less prepared to cope with excessive temperatures. Young children, older adults, people with medical conditions... are more vulnerable than others to heat-related illness. The share of the U.S. population composed of adults over age 65 is currently 12%, but is projected to grow to 21% by 2050, leading to a larger vulnerable population.

Impacts from Extreme Weather Events

The frequency and intensity of extreme precipitation events is projected to increase in some locations, as is the severity (wind speeds and rain) of tropical storms. These extreme weather events could cause injuries and, in some cases, death. As with heat waves, the people most at risk include young children, older adults, people with medical conditions, and the poor. Extreme events can also indirectly threaten human health in a number of ways. For example, extreme events can:

> Reduce the availability of fresh food and water. Interrupt communication, utility, and health care services. Contribute to carbon monoxide poisoning from portable electric generators used during and after storms. Increase stomach and intestinal illness among evacuees. Contribute to mental health impacts such as depression and post-traumatic stress disorder (PTSD).

Impacts from Reduced Air Quality

Despite significant improvements in U.S. air quality since the 1970s, as of 2008 more than 126 million Americans lived in counties that did not meet national air quality standards.

Increases in Ozone: Scientists project that warmer temperatures from climate change will increase the frequency of days with unhealthy levels of ground-level ozone, a harmful air pollutant, and a component in smog.

* Ground-level ozone can damage lung tissue and can reduce lung function and inflame airways. This can increase respiratory symptoms and aggravate asthma or other lung diseases. It is especially harmful to children, older adults, outdoor workers, and those with asthma and other chronic lung diseases.

* Ozone exposure is associated with increased susceptibility to respiratory infections, medication use, doctor visits, and emergency department visits and hospital admissions for individuals with lung disease. Some studies suggest that ozone may increase the risk of premature mortality, and possibly even the development of asthma.

* Because warm, stagnant air tends to increase the formation of ozone, climate change is likely to increase levels of ozone in already-polluted areas of the United States and increase the number of days with poor air quality.

Changes in Fine Particulate Matter

Particulate matter is the term for a category of extremely small particles and liquid droplets suspended in the atmosphere. Fine particles include particles smaller than 2.5 micrometers (about one ten-thousandth of an inch). These particles may be formed in the atmosphere from chemical reactions of gases such as sulfur dioxide, nitrogen dioxide, and volatile organic compounds.

* Inhaling fine particles can lead to a broad range of adverse health effects, including premature mortality, aggravation of cardiovascular and respiratory disease, development of chronic lung disease, exacerbation of asthma, and decreased lung function growth in children.

* Sources of fine particle pollution include power plants, gasoline and diesel engines, wood combustion, high-temperature industrial processes such as smelters and steel mills, and forest fires.

Changes in Allergens

Climate change may affect allergies and respiratory health. The spring pollen season is already occurring earlier in the United States due to climate change. The length of the season may also have increased. In addition, climate change may facilitate the spread of ragweed, an invasive plant with very allergenic pollen. Tests on ragweed show that increasing carbon dioxide concentrations and temperatures would increase the amount and timing of ragweed pollen production.

Impacts from Climate-Sensitive Diseases

Changes in climate may enhance the spread of some diseases. [1] Disease-causing agents, called pathogens, can be transmitted through food, water, and animals such as deer, birds, mice, and insects. Climate change could affect all of these transmitters.

Food-borne Diseases

* Higher air temperatures can increase cases of salmonella and other bacteria-related food poisoning because bacteria grow more rapidly in warm environments. These diseases can cause gastrointestinal distress and, in severe cases, death.

* Flooding and heavy rainfall can cause overflows from sewage treatment plants into fresh water sources. Overflows could contaminate certain food crops with pathogen-containing feces.

Water-borne Diseases

* Heavy rainfall or flooding can increase water-borne parasites such as Cryptosporidium and Giardia that are sometimes found in drinking water. These parasites can cause gastrointestinal distress and in severe cases, death.

* Heavy rainfall events cause stormwater runoff that may contaminate water bodies used for recreation (such as lakes and beaches) with other bacteria. The most common illness contracted from contamination at beaches is gastroenteritis, an inflammation of the stomach and the intestines that can cause vomiting, headaches, and fever.

Animal-borne Diseases

The geographic range of ticks that carry Lyme disease is limited by temperature. As air temperatures rise, the range of these ticks is likely to expand northward. Typical symptoms of Lyme disease include fever, headache, fatigue, and a characteristic skin rash. In 2002, a new strain of West Nile virus, which can cause serious, life-altering disease, emerged in the United States. Higher temperatures are favorable to the survival of this new strain.

Other Health Linkages

Other linkages exist between climate change and human health. Changes in temperature and precipitation, as well as droughts and floods, will likely affect agricultural yields and production. In some regions of the world, these impacts may compromise food security and threaten human health through malnutrition, the spread of infectious diseases, and food poisoning. The worst of these effects are projected to occur in developing countries, among vulnerable populations.

Although the impacts of climate change have the potential to affect human health in the United States and around the world, there is a lot we can do to prepare for and adapt to these changes. Learn about how we can adapt to climate impacts on health.

The Psychological Effects of
GLOBAL WARMING
on the United States

Excerpts from the Executive Summary of a 2011 report prepared
by Dr. Lise Van Susteren, MD, with Dr. Kevin Coyle, JD

**Global warming…
in the coming years…
will foster public trauma,
depression, violence,
alienation, substance
abuse, suicide, psychotic
episodes, post-traumatic
stress disorders and
many other mental
health-related conditions.**

THE EXTREME AND SOMETIMES VIOLENT weather of the summer of 2011 can offer valuable insights into how a warming climate will affect the people in the United States and other parts of the world. The news headlines included: a worsening Texas drought, record heat in the eastern states, a rise in heat-related deaths in many U.S. cities; violent floods in the East and Midwest; an expanded range and season for some of the worst tornados on record and more.

These same headlines included the seemingly unrelated famine and refugee tragedy in Somalia, a rise in mental health difficulties among service men and women returning from war, and anomalous weather conditions and disease outbreaks in many parts of the world.

Climate scientists have begun to empirically link 2011's extreme weather events and natural disasters to climate change and report that these are representative of what science predicts the world will look like with more warming. The physical and economic harm caused by such events is evident, but what will be the toll on the public's mental health?

To those who would deny, dismiss or just fail to envision the psychological impacts global warming, we urge you to take a deeper look. We may not currently be thinking about how heavy the toll on our psyche will be, but, before long, we will know only too well. A warming climate will cause many people, tens of millions, to hurt profoundly.

Global warming from increased greenhouse gases in our atmosphere is leading to a spiral of worsening conditions that will include extreme and sometime violent weather. What we are already seeing is alarming indeed: in 2011 alone we faced devastating droughts, raging wildfires, record breaking snowstorms and rainfalls, stunning floods in the East and Midwest, higher temperatures and more frequent 100 degree days in more cities than we have ever known - with a commensurate rise in heat related deaths, an expanded range and season for some of the worst tornadoes on record, and the most costly hurricane in our history.

The U.S. mental health care system is only minimally prepared to address the effects of global climate change-related disasters and incidents.

In November of 2011, the U.N. sponsored Intergovernmental Panel on Climate Change confirmed this in a report entitled: *Special Report on Managing the Risks of Extreme Events and Disasters to Advance Climate Change Adaptation.* The report finds that changes in weather, due to climate warming, will be felt everywhere in the world. The physical and economic destruction surely boggles the mind but what is not being addressed are the human psychological consequences of all this devastation.

To begin with, the incidences of mental and social disorders will rise steeply. These will include depressive and anxiety disorders, post traumatic stress disorders, substance abuse, suicides, and widespread outbreaks of violence. Children, the poor, the elderly, and those with existing mental health disorders are especially vulnerable and will be hardest hit. At roughly 150 million people, these groups represent about one half of the American public.

The American mental health community, counselors, trauma specialists and first responders are not even close to being prepared to handle scale and intensity of impacts that will arise from the harsher conditions and disasters that global warming will unleash. It is not that we haven't experienced natural disasters before, but the scientific data show that what lies ahead will be bigger, more frequent, and more extreme than we have ever known.

There are even broader implications, many of them beyond our shores. As climate related disasters and burdens spread across the world, the U.S. military will increasingly be called upon to help keep order. Service members will be faced with stressful, even horrifying conditions. They will see people - the young, the old, the innocent – suffer terribly. Back home their families will experience the ripple effects, suffering vicariously and experiencing their own disruptions in finances, relationships and child-rearing. There will be the disorders from the immediate trauma, and in some cases *chronic* psychological disorders will follow.

Another major problem for the military is a high rate of active service member suicide. Even though the numbers have recently declined after reaching a high of nearly double the rate of the civilian population, the problem persists. While suicide is the result of many complex factors, the linkage to global warming with respect to military personnel must be acknowledged. Burning fossil fuels for energy means depending on foreign areas where those supplies are most abundant. To the U.S. military this can mean sending young people into battle to protect our energy sources or to calm related unrest. Our service members will recognize that their own lives and limbs were sacrificed even though alternate renewable sources of energy could be more available. Our national need to put these young people in harm's way would also decline if we were simply more energy efficient. How will we answer these service members' questions about why we didn't work harder at fixing this problem?

It is not a matter of whether these problems will occur, but rather how frequently and with what intensity.

Moreover, the United States is increasingly disliked, worldwide, as a global warming villain. Though representing less than 5 percent of the world's population, the U.S. emits about 25 percent of the world's green house gases. As the link between climate disasters in other countries and the production of green house gases in the

U.S. becomes clearer, Americans will be blamed for inflicting harm on other countries. Critics may point to emissions from China (now surpassing the U.S.) and India as reasons why the U.S. can "share the blame," but our per capita emissions are second to none.

Alarmingly, our perceived indifference is already the subject of rallying cries against us. It is used by leaders of terrorist groups, for example, as a tool to recruit new members. The President of one African country hit hard by drought linked to climate change addressed countries emitting high levels of green house gases: "We have a message here to tell these countries, that you are causing aggression to us by causing global warming." The President of Bolivia, faced with unprecedented flooding from heavy rains, threatened to sue the U.S. in international court.

The U.S. Department of Defense predicts that events linked to climate change, such as crop failures, water shortages, disease outbreaks, and more will soon be the leading cause of world turmoil. Unstable states, faced with these stressors, are at risk of slipping into chaos, and failing. This paves the way for takeovers by groups hostile to the U.S. and is a growing reality widely feared by our military.

Some 50 million elderly people, and America's 35 million low-income people will suffer a disproportionate amount of physical and psychological stress.

The economic costs of climate change will be high by any measure. But its specific effect on U.S. mental health, societal well being and productivity will increase current U.S. expend- itures on mental health services adding to our current $300 billion annual burden. Incredibly this probable cost is overlooked in today's national public health debate and environmental discussions. The U.S. mental health care system is not prepared to address the full effects of global warming-related disasters and incidents. A comprehensive assessment of what will be required begs to be undertaken. Training health care providers and first responders to address the

large-scale mental distress arising from the emergencies that are coming is imperative. Timely interventions may keep some early injuries from developing into chronic, long-term conditions.

SUMMARY OF FINDINGS

An estimated 200 million Americans will be exposed to serious psychological distress from climate related events and incidents:

The severity of symptoms will vary, but in many instance the distress will be great.

In the coming years, a majority of Americans will experience direct adverse effects from the impacts of global warming. Natural disasters and extreme weather events will strike many places that are densely populated: 50 percent of Americans live in coastal regions exposed to storms and sea level rise, 70 percent of Americans live in cities prone to heat waves; major inland cities lie along rivers that will swell to record heights, and the fastest growing part of the nation is the increasingly arid West.

Climate change will become a top-of-mind worry in the future:

Some Americans already are or will soon experience anxiety about global warming and its effects on us, our loved ones, our ecosystems, and our lifestyles. This anxiety will increase as reports of the gravity of our condition become more clear and stark. ...

Major segments of U.S. society are more psychologically vulnerable now:

• Children: America's 70 million children will not only suffer long term effects from climate change but will also experience acute reactions to natural disasters and extreme weather events.

Some children are already anxious about global warming and begin to obsess (understandably) about the future, unmoved by the small reassurances adults may attempt to put forth. In the first known "climate change delusion" a depressed 17 year old boy was hospitalized for refusing to drink water out of fear it would cause many more deaths in drought ridden Australia. The doctor who treated him has seen other children suffering from climate related anxiety disorders.

The mental health care system of the U.S. is not prepared to handle the wide-spread psychological stresses of climate change:

While the U.S. mental health care professions are coming to recognize and address the larger scale perils associated with climate change, no comprehensive strategies are in place to cope with the full psychological and public mental health implications. Given the magnitude of the impacts and the rate at which the world is changing, a campaign focused on what this segment of the U.S. mental health service community can do to help is certainly needed....

There is also low first responder preparedness:

Due to the number of emergency situations in which global climate change and mental health issues will be connected, first responders will need additional education and training to handle the immediate psychological trauma and symptoms of climate disaster victims. Such training will support rescue operations, triage decisions, application of medications, patient safety and more....

Some climate change-related conditions and their psychological effects merit specific preparation:

• **Summer heat waves:** the physical distress arising from prolonged heat waves is well known. What is not widely known is the psychological distress that is caused by higher temperatures, and, in particular, the relationship between rising temperature and aggression. Research from Iowa State shows that, as the temperature rises, so does the incidence of violence. (DeLisi, 2010)

• **Coastal and river flooding:** the direct adverse effects of flooding are obvious, but these weather and climate related events are especially likely to lead to psychic injury from the stress of displace-ment, loss of possessions (including pets), and uncertainty about interim and future housing and employment.

• **High impact and more intense storms:** the far-reaching consequences of destructive weather saw its prototype in Hurricane Katrina. The Hurricane scattered residents of New Orleans all across the U.S. It shattered a culture, broke up families, spiked outbursts of outrage and blame at a government that was slow to respond, and lead to a jump in violence in at least one city that took them in (Houston). Six years later New Orleans has yet to fully recover, and many of the victims have experienced post-incident stress and post-traumatic stress disorders (PTSD)....

• **Severe drought and reduced snow pack:** the unrelenting day by day despair of watching and waiting for water that doesn't come will have a singularly damaging impact on the psyche of the people who have depended on Mother Nature's rainfall for their livelihood....

• **Increased large-scale wildfires:** raging wildfires are incredibly dangerous and have a particularly savage effect on our psyches by devastating landscapes, wiping out homes and possessions, incinerating wildlife and clogging the air with pollutants that sicken people locally and can travel hundreds of miles to sicken people at a distance. Persistent psychological stress is common, with anxiety reactions recurring....

• **New Disease threats:** higher temperatures favor the formation of ozone which triggers asthma attacks. Anyone who has asthma and parents of children with asthma are familiar with the fears this illness engenders. People die from untreated asthma. Many other fears linked to disease are harder to "nail down." As malaria and dengue fever and other infectious diseases march northward due to warmer temperatures, inchoate fears of threat and vulnerability drift into people's consciousness. This will be compounded by a growing number of sensational media reports tied to disease outbreaks and public health warnings....

We are witnessing an unraveling of climate stability, and therefore human stability, and are seeing physical changes that are unprecedented in all of history.

THE WHITE HOUSE

* * * * * WASHINGTON * * * * *

Office of the Press Secretary

For Immediate Release

May 06, 2014

(abridged)

FACT SHEET

What Climate Change Means for Regions across America and Major Sectors of the Economy

"...Science, accumulated and reviewed over decades, tells us that our planet is changing in ways that will have profound impacts on all of humankind... Those who are already feeling the effects of climate change don't have time to deny it—they're busy dealing with it."

– President Barack Obama
Remarks at Georgetown University
June 25, 2013

TODAY, DELIVERING ON A MAJOR COMMITMENT IN THE PRESIDENT'S Climate Action Plan, the Obama Administration is unveiling the third U.S. National Climate Assessment—the most comprehensive scientific assessment ever generated of climate change and its impacts across every region of America and major sectors of the U.S. economy.

The findings in this National Climate Assessment underscore the need for urgent action to combat the threats from climate change, protect American citizens and communities today, and build a sustainable future for our kids and grandkids.

Developed over four years by hundreds of the Nation's top climate scientists and technical experts—and informed by thousands of inputs from the public and outside organizations gathered through town hall meetings, public-comment opportunities, and technical workshops across the country, the third National Climate Assessment represents the most authoritative and comprehensive knowledge base about how climate change is affecting America now, and what's likely to come over the next century.

And, for the first time, to ensure that American citizens, communities, businesses, and decision makers have easy access to scientific information about climate change impacts that are most relevant to them, the U.S. National Climate Assessment is being released in an interactive, mobile-device-friendly, digital format on www.globalchange.gov.

Today's announcement is a key deliverable of the Climate Action Plan launched by President Obama last June—which lays out concrete steps to cut carbon pollution, prepare America's communities for climate-change impacts, and lead international efforts to address this global challenge. The Plan acknowledges that even as we act to reduce the greenhouse-gas pollution that is driving climate change, we must also empower the Nation's communities, businesses, and individual citizens with the information they need to cope with the changes in climate that are already underway.

Climate-Change Impacts in Regions across America:

Northeast – *Maine, New Hampshire, Vermont, Massachusetts, Rhode Island, Connecticut, New York, New Jersey, Delaware, Pennsylvania, Maryland, and District of Columbia:* The high-density urban coastal corridor from Washington, DC, to Boston is one of the most developed environments in the world, containing a massive, complex, network of supporting infrastructure.... Communities in the Northeast "are affected by heat waves, more extreme precipitation events, and coastal flooding due to sea level rise and storm surge."

Southeast and Caribbean –*Virginia, W. Virginia, Kentucky, Tennessee, Georgia, Alabama, Arkansas, S. Carolina, N. Carolina, Mississippi, Florida, Louisiana, and the Caribbean Islands*: The Southeast and Caribbean region "is home to more than 80 million people and some of the fastest-growing metropolitan areas.... Decreased water availability, exacerbated by population growth and land-use change, causes increased competition for water.... There are also increased risks associated with extreme events such as hurricanes."

Midwest – *Minnesota, Michigan, Iowa, Indiana, Ohio, Missouri, Illinois, and Wisconsin:* "The Midwest's agricultural lands, forests, Great Lakes, industrial activities, and cities are all vulnerable to climate variability and climate change.... Longer growing seasons and rising carbon dioxide levels increase yields of some crops, although these benefits have already been offset in some instances by occurrence of extreme events such as heat waves, droughts, and floods."

Great Plains – *Wyoming, N. Dakota, S. Dakota, Montana, Nebraska, Kansas, Oklahoma, and Texas*: The Great Plains region "experiences multiple climate and weather hazards, including floods, droughts, severe storms, tornadoes, and winter storms. In much of the Great Plains, too little precipitation falls to replace that [which is] needed.... These conditions already stress communities and cause billions of dollars in damage. Climate change will add to both stress and costs." "Rising temperatures lead to increased demand for water and energy and impacts on agricultural practices."

Southwest – *California, Nevada, Arizona, New Mexico, Utah, and Colorado*: "The Southwest is the hottest and driest region in the United States. Climate changes pose challenges for an already parched region that is expected to get hotter and, in its southern half, significantly drier. Increased heat and changes to rain and snowpack will send ripple effects throughout the region... and its critical agriculture sector." "Drought and increased warming foster wildfires and increased competition for scarce water resources for people and ecosystems."

Northwest – *Idaho, Oregon, and Washington:* "The Northwest's economy, infrastructure, public health, and agriculture sectors face important climate change related risks. Impacts on infrastructure, natural systems, human health, and economic sectors, combined with issues of social and ecological vulnerability, will unfold quite differently in largely natural areas, like the Cascade

Range, than in urban areas like Seattle and Portland or region's many Native American Tribes. Changes in the timing of streamflow related to earlier snowmelt reduce the supply of water in summer, causing far-reaching ecological and socioeconomic consequences."

Alaska: "Over the past 60 years, Alaska has warmed more than twice as rapidly as the rest of the United States…The state's largest industries, energy production, mining, and fishing—are all affected by climate change." "Rapidly receding summer sea ice, shrinking glaciers, and thawing permafrost cause damage to infrastructure and major changes to ecosystems. Impacts on Alaska Native communities increase."

Hawaii and Pacific Islands: The U.S. Pacific Islands region "includes more than 2,000 islands spanning millions of square miles of ocean. Rising air and ocean temperatures, shifting rainfall patterns, changing frequencies and intensities of storms and drought, decreasing streamflows, rising sea levels, and changing ocean chemistry threaten the sustainability of globally important and diverse ecosystems…as well as local communities, livelihoods, and cultures." "Increasingly constrained freshwater supplies, coupled with increased temperatures, stress both people and ecosystems and decrease food and water security."

Coasts: "More than 50% of Americans... live in coastal counties.... Humans have heavily altered the coastal environment through development, changes in land use, and overexploitation.... Now, the changing climate is imposing additional stresses.... Coastal lifelines, such as water supply infrastructure and evacuation routes are increasingly vulnerable to higher sea levels and storm surges, inland flooding, and other climate-related changes."

Climate-Change Impacts on Key Sectors of Society and the U.S. Economy

Health: "Climate change threatens human health and well-being including through impacts from increased extreme weather events, wildfire, decreased air quality, threats to mental health, and illnesses transmitted by food, water, and disease carriers such as mosquitoes and ticks. Some of these health impacts are already underway in the United States. Climate change will... amplify existing health threats.... Certain people and communities are especially vulnerable, including children, the elderly, the sick, the poor, and some communities of color. Public health actions, especially preparedness and prevention, can do much to protect people from some of the impacts of climate change. Early action provides the largest health benefits."

Transportation: "The impacts from sea level rise and storm surge, extreme weather events, higher temperatures and heat waves, precipitation changes, Arctic warming, and other climatic conditions are affecting the reliability and capacity of the U.S. transportation system in many ways. Sea level rise, coupled with storm surge, will... increase the risk of coastal impacts on transportation infrastructure, including temporary and permanent flooding of airports, ports and harbors, roads, rail lines, tunnels, and bridges. Extreme weather events currently disrupt transportation networks in all areas of the country; projections indicate that such disruptions will increase. Climate change impacts will increase the costs to the Nation's transportation systems and their users, but these impacts can be reduced through rerouting, mode change, and a wide range of adaptive actions."

Energy: "Extreme weather events are affecting energy production and delivery, facilities causing supply disruptions... and affecting infrastructure that depends on energy supply. The frequency and intensity of extreme weather events are expected to change. Higher summer temperatures will increase electricity use, causing higher summer peak loads, while warmer winters will decrease demands for heating. Changes in water availability... will constrain energy production. In the longer term, sea level rise, extreme storm surge events, and high tides will affect coastal facilities and infrastructure on which many energy systems, markets, and consumers depend. As new energy technologies occur, future energy systems will differ from today's in uncertain ways. Depending on the changes in the energy mix, climate change will introduce new risks as well as new opportunities."

Water: "Climate change affects water demand and the ways water is used within and across regions and economic sectors. The Southwest, Great Plains, and Southeast are vulnerable to changes in water supply and demand. Changes in precipitation and runoff, combined with changes in consumption and withdrawal, have reduced surface and groundwater supplies.... These trends are expected to continue, increasing the likelihood of water shortages.... Increasing flooding risk affects human safety and health, property, infrastructure, economies, and ecology in many basins across the United State. Increasing resilience and enhancing adaptive capacity provide opportunities to strengthen water resources management and plan for climate-change impacts."

Agriculture: "Climate disruptions to agriculture have been increasing and are projected to become more severe.... Some areas are already experiencing climate-related disruptions, particularly due to extreme weather events. While some U.S. regions and some types of agricultural production will be relatively resilient over the next 25 years, others will suffer from stresses due to extreme heat, drought, disease, and heavy downpours. From mid-century on, climate change is projected to have more negative impacts on crops and livestock – a trend that could diminish the security of our food supply... Climate change effects on agriculture will have consequences for food security, through changes in crop yields and food prices and effects on food processing, storage, transportation, and retailing. Adaptation measures can help reduce some of these impacts."

Ecosystems: "Ecosystems and the benefits they provide to society are being affected by climate change. The capacity of ecosystems to buffer extreme events like fires, floods, and severe storms is being overwhelmed. Climate change impacts on biodiversity are already observed in alteration of the timing of critical biological events such as spring bud burst, and range shifts of many species. In the longer term, there is an increased risk of species extinction. Events such as droughts, floods, wildfires, and pest outbreaks associated with climate change (for example, bark beetles in the West) are already disrupting ecosystems. These changes limit the capacity of ecosystems, such as forests, barrier beaches, and wetlands, to play important roles in reducing the impacts of extreme events on infrastructure, human communities, and other valued resources... Whole-system management is more effective than focusing on one species at a time, and can help reduce the harm to wildlife, natural assets, and human well-being that climate disruption might cause."

Oceans: "Ocean waters are becoming warmer and more acidic, broadly affecting ocean circulation, chemistry, ecosystems, and marine life. More acidic waters inhibit the formation of shells, skeletons, and coral reefs. Warmer waters harm coral reefs and alter the distribution, abundance, and productivity of many marine species. The rising temperature and changing chemistry of ocean water combine with other stresses, such as overfishing and coastal and marine pollution, to alter marine-based food production and harm fishing communities... In response to observed and projected climate impacts, some existing ocean policies, practices, and management efforts are incorporating climate change impacts. These initiatives can serve as models for other efforts and ultimately enable people and communities to adapt to changing ocean conditions."

Climate Trends in America

Temperature: "U.S. average temperature has increased by 1.3° F to 1.9° F since record keeping began in 1895; most of this increase has occurred since 1970. The most recent decade was the Nation's warmest on record. Temperatures in the United States are expected to continue to rise. Because human-induced warming is superimposed on a naturally varying climate, the temperature rise has not been, and will not be, uniform or smooth across the country or over time."

Extreme Weather: "There have been changes in some types of extreme weather events over the last several decades. Heat waves have become more frequent and intense, especially in the West. Cold waves have become less frequent and intense across the Nation. There have been regional trends in floods and droughts. Droughts in the Southwest and heat waves everywhere are projected to become more intense, and cold waves less intense everywhere."

Hurricanes: "The intensity, frequency, and duration of North Atlantic hurricanes, as well as the frequency of the strongest (Category 4 and 5) hurricanes, have increased since the early 1980s. The contributions of human and natural causes to these increases are still uncertain. Hurricane-associated storm intensity and rainfall rates are projected to increase as climate continues to warm."

Severe Storms: "Winter storms have increased in frequency and intensity since the 1950s, and their tracks have shifted northward over the United States. Other trends in severe storms, including the intensity and frequency of tornadoes, hail, and damaging thunderstorm winds, are uncertain and are being studied intensively." (*NCA Highlights: Climate Trends*)

Precipitation: "Average U.S. precipitation has increased since 1900, but some areas have had increases greater than the national average, and some areas have had decreases. More winter and spring precipitation is projected for the northern United States, and less for the Southwest, over this century."

Heavy Downpours: "Heavy downpours are increasing nationally, especially over the last three to five decades. Largest increases are in the Midwest and Northeast. Increases in the frequency and intensity of extreme precipitation events are projected for all U.S. regions."

Frost-free Season: "The length of the frost-free season (and the corresponding growing season) has been increasing nationally since the 1980s, with the largest increases occurring in the western United States, affecting ecosystems and agriculture. Across the United States, the growing season is projected to continue to lengthen."

Ice Melt: "Rising temperatures are reducing ice volume and surface extent on land, lakes, and sea. This loss of ice is expected to continue. The Arctic Ocean is expected to become essentially ice free in summer before mid-century."

Sea Level: "Global sea level has risen by about 8 inches since reliable record keeping began in 1880. It is projected to rise another 1 to 4 feet by 2100."

Ocean Acidification: "The oceans are currently absorbing about a quarter of the carbon dioxide emitted to the atmosphere annually and are becoming more acidic as a result, leading to concerns about intensifying impacts on marine ecosystems."

Union of Concerned Scientists

Science for a healthy planet and a safer world

The Scientific Consensus on Global Warming

Scientific societies and scientists have released statements and studies showing the growing consensus on climate change science. A common objection to taking action to reduce our heat-trapping emissions has been uncertainty within the scientific community on whether or not global warming is happening and if it is caused by humans. However, there is now an overwhelming scientific consensus that global warming is indeed happening and humans are contributing to it. Below are links to documents and statements attesting to this consensus.

Scientific Societies

American Meteorological Society

Climate Change: An Information Statement of the American Meteorological Society

"Indeed, strong observational evidence and results from modeling studies indicate that, at least over the last 50 years, human activities are a major contributor to climate change." (February 2007)

American Physical Society

Statement on Climate Change

"The evidence is incontrovertible: Global warming is occurring. If no mitigating actions are taken, significant disruptions in the Earth's physical and ecological systems, social systems, security and human health are likely to occur. We must reduce emissions of greenhouse gases beginning now." (November 2007)

American Geophysical Union

Human Impacts on Climate

"The Earth's climate is now clearly out of balance and is warming. Many components of the climate system — including the temperatures of the atmosphere, land and ocean, the extent of sea ice and mountain glaciers, the sea level, the distribution of precipitation, and the length of seasons—are now changing at rates and in patterns that are not natural and are best explained by the increased atmospheric abundances of greenhouse gases and aerosols generated by human activity during the 20th century." (Adopted, December 2003, Revised and Reaffirmed. December 2007)

American Association for the Advancement of Science

AAAS Board Statement on Climate Change

"The scientific evidence is clear: global climate change caused by human activities is occurring now, and it is a growing threat to society." (December 2006)

Geological Society of America

Global Climate Change

"The Geological Society of America (GSA) supports the scientific conclusions that Earth's climate is changing; the climate changes are due in part to human activities; and the probable consequences of the climate changes will be significant and blind to geopolitical boundaries." (October 2006, reaffirmed, March, 2013)

American Chemical Society

Statement on Global Climate Change

"There is now general agreement among scientific experts that the recent warming trend is real (and particularly strong within the past 20 years), that most of the observed warming is likely due to increased atmospheric greenhouse gas concentrations, and that climate change could have serious adverse effects by the end of this century." (July 2004)

U.S. National Academy of Sciences

Understanding and Responding to Climate Change

"The scientific understanding of climate change is now sufficiently clear to justify taking steps to reduce the amount of greenhouse gases in the atmosphere." (2005)

International Academies

Joint science academies' statement: Global response to climate change

"Climate change is real. There will always be uncertainty in understanding a system as complex as the world's climate. However there is now strong evidence that significant global warming is occurring." (2005, 11 national academies of science)

"Despite increasing consensus on the science underpinning predictions of global climate change, doubts have been expressed recently about the need to mitigate the risks posed by global climate change. We do not consider such doubts justified." (2001, 16 national academies of science)

The Smithsonian Institution

NEWS DESK: News Room of the Smithsonian

Smithsonian Statement on Climate Change

October 2, 2014
abridged

Rapid and long-lasting climate change is a topic of growing concern as the world looks to the future. Scientists, engineers and planners are seeking to understand the impact of new climate patterns, working to prepare our cities against the perils of rising storms and anticipating threats to our food, water supplies and national security. Scientific evidence has demonstrated that the global climate is warming as a result of increasing levels of atmospheric greenhouse gases generated by human activities. A pressing need exists for information that will improve our understanding of climate trends, determine the causes of the changes that are occurring and decrease the risks posed to humans and nature.

Climate change is not new to the Smithsonian — our scholars have investigated the effects of climate change on natural systems for more than 160 years. We look at processes that occurred millions of years ago alongside developments taking place in today's climate system. ...

The urgency of climate change requires that we boost and expand our efforts to increase public knowledge....

The Smithsonian will continue, as it has for more than a century and a half, to produce basic scientific information about climate change and to explore the cultural and historical significance of these changes. The urgency of climate change requires that we boost and expand our efforts to increase public knowledge and that we inspire others through education and by example. We live in what has come to be called the Anthropocene, or "The Age of Humans." The Smithsonian is committed to helping our society make the wise choices needed to ensure that future generations inherit a diverse world that sustains our natural environments and our cultures for centuries to come.

# # #

Why the Denial about Climate Change?

The Dilemma

While the science of climate change is becoming clearer, and while we have just ended the hottest decade and what appears to be the hottest year on record, we still do not have large numbers of people awakening to the urgency of climate change as a serious issue. Examination finds a set of economic calculations at work.

The Economic Issues Surrounding Climate Change

The dilemma is first economic. The problem is not a lack of scientific understanding. It is precisely the opposite. Global climate change represents a threat to many perspectives that corporations hold dear. Climate change will upend support for free trade because climate change implies a need to localize economies. Globalization and free trade would end because we would have to address the moral issue of inequality. We would also need to subsidize the global South because the climate crisis was born in the North, but it is most acutely impacting in the global South. Justice will require the payment of reparations.

Increased Regulation

Corporations would have to be regulated because stronger controls over the economy would require more intense levels of intervention in commerce. Fossil fuels would be charged a fee and clean energy sources subsidized.

Increased Federal and International Control

A stronger United Nations with enforcement would be necessary as no nation can hold off climate change alone. Renewable energy sources would have to be subsidized to break the grip of dirty fossil fuels. This would clash with the political ideology of the right.

Denial by so-called Green Groups

It is not just those on the political right who are in denial. Some environmental groups share culpability in that they propose tiny steps when our whole cultural outlook is being indicted by climate change. Some green groups suggest that we might turn to "green capitalism," and that we need only to change few light bulbs and drive less. Anyone who says this does not yet grasp the challenge which society is facing. Climate change will change our whole way of life and we need to begin by examining our own lives. Local food, for instance, requires only one-tenth of the energy as imported food.

Quick Conclusion

It is easier to deny climate science than to deny the corporate ideological agenda. What is at stake is an entire philosophical, political and economic worldview. Climate change will also imply a renunciation of consumerism. It supports a more frugal, thoughtful and sustainable and more local mode of living. We are still only recognizing the beginning of the massive changes that will be necessary to successfully address global climate change.

What You and Your Congregation Can Do

GLOBAL CLIMATE CHANGE IS A MONSTROUSLY HUGE ISSUE AND IT SEEMS hard for one person to make a difference. Yet because we are dealing with a global issue caused by the actions of millions upon millions of people, most of whom believe that they are acting as good citizens, going about their lives doing proper and honorable activities, we have to begin the process of change by acting within our own spheres. This gives our actions the backbone of integrity and allows us to speak with strength born of living what we advocate. This is central to the moral argument on climate change. We have to become in our own lives and behavior answers and solutions to the climate problem.

Besides, we cannot expect government to act if we, the people, fail to demonstrate in our own lives, a willingness to correct our errors of behavior. For most of us, we just didn't know how fossil fuels could pollute the entire planet. Now we know. And now we no longer have any excuse for continuing in habits that pollute the earth and defile the planetary commons. When we did not know that we are harming the earth, we could be absolved of polluting its atmosphere. But now we know. To continue in ways that defile the earth is to be guilty of willful participation in the forces that degrade and defile and ultimately sicken the world. For Christians, this is not just a small sin. As the author of the Book of Revelation reveals, *"Those who destroy the earth, God will destroy"* (Revelation 11:18). Can we imagine a stronger biblical admonition than that?

Scripture makes it clear that people should not only care for the earth, but should take deliberate steps to correct whatever defiles the earth. The guidances on care for the earth become easier because we also discern that for those who protect what is first God's, the blessings of God come upon them as they serve and protect and what God has placed into our care and keeping.

The following principles and recommended actions can serve as a beginning for addressing the serious issues facing each one of us and all the people and living things of the good Earth.

Guiding Principles for Addressing Climate Change

The following themes offer beginning guidance to help you fight climate change in your home, business or house of worship. Some of these actions seem simple, and they are, but they are important to society's ability to correct the root causes of climate change. Make every effort to apply these guidelines as these will put you on the path toward becoming a champion at healing the earth.

◆ PRACTICE SUSTAINABILITY

Sustainability is the practice of the Three R's: Reduce, Reuse, and Recycle. The trash of consumerism contributes to climate change in many ways. Plastics and other "disposable" products like Styrofoam release greenhouse gasses in their production and do not disintegrate over time. Food waste builds up in trash heaps to create methane, a strong greenhouse gas. Recycling plastic bottles (and glass and paper) reduces one's "carbon footprint" and prevents more new resources from being used to create products. Composting your food rather than throwing it into the garbage prevents the buildup of methane in trash heaps and further reduces your contribution to greenhouse gases. Through your power as a consumer, you can reduce the products you buy, thus decreasing demand for products that create greenhouse gas emissions and contribute to climate change.

◆ CHOOSE CLIMATE FRIENDLY FOODS

Agriculture and the food industry contributes to climate change in many ways. The fertilizers used in growing crops, and the energy used in the transportation of food and the tractors and other equipment that process food all contribute to greenhouse gas emissions. Additionally some foods contribute more to greenhouse gas emissions than others. An easy way to choose foods that are low in greenhouse gases is to select local, organic whole food options which statistically release fewer greenhouse gases in the growing, fertilizing and transporting process and to eat lower on the food chain. (This does not mean that everyone needs to become vegetarian, although that does not hurt and it is healthier, both for the person and the planet.)

◆ INSTALL RENEWABLE ENERGY

Renewable energy means that it can be used forever and it will not release greenhouse gas emissions. Many times renewable options can be installed in your home so that the electrical energy you use has zero contribution to climate change. The most common way to achieve this is by installing solar power on your house, or near your house of worship, a process that can be very affordable when bought through the right organization. Other options include wind power and geothermal power. Although not common in the United States, many farmers in Germany put wind turbines in their fields so that the machinery they use also becomes renewable.

◆ DRIVE LESS

Driving is one of the most direct ways the everyday person contributes to greenhouse gas emissions. But for many who have long commutes or drive their children across town to school each day, the mileage cannot be reduced. Instead of reducing your mileage, look for ways to reduce the amount of fuel you use. Public transportation, joining a carpool group, putting your child on a school bus, or investing in a more fuel-efficient car are great ways to start.

◆ PRACTICE ENERGY EFFICIENCY

Using less energy in your home, office, and house of worship combats climate change in two ways: As consumers acting together, you can impact greenhouse gas emissions from fossil fuel companies by reducing your demand. You can also reduce your own "carbon footprint" by using less energy. There are many ways to reduce energy use in your home. Some of the easiest include installing florescent or LED light bulbs, turning out the light when not in a room, using low energy appliances.

◆ DIVEST FROM FOSSIL FUELS

A campaign is emerging that emphasizes the "mom and pop" strategy; that is, "move our money" and "protect our planet." This demonstrates that we will not support efforts that cause harm to the planet. Besides fossil fuel investments are not safe in the long term, as declining fossil fuel reserves will in the future lead to declines in share prices. "Stranded assets" is the term for investments that lose value because the investments are no longer of much value.

◆ VOTE THE CLIMATE

Prioritize candidates in your choice of elected officials that acknowledge the seriousness of our climate predicament and that listen to sound science and the voice of religion on climate. There is no issue of greater significance for both our nation's and our children's future than our choices of energy as these relate to how we will power the future.

◆ GET INVOLVED

Improving education and awareness about climate change is important. A person who does not understand the causes and implications of climate change will not understand the urgency to fight it. Explain what is happening. Volunteer to work on a climate change campaign. If one does not exist where you live, start on in your house of worship, or contribute to an organization that is active. Form groups that teach the seriousness of climate change. Work with others and form networks to support local conferences and ways to improve awareness about the urgent issue of global climate change. Do all that you can to help others grasp the implications for the future and for our children of climate change. Encourage clergy to talk about climate change and offer ways for it to be addressed. Amplify the voice of religion as it is increasingly speaking with one voice to the world on the necessity to address those actions causing climate change.

Actions to Reduce Carbon Emissions

Here are some specific first steps that you can take to address climate change:

At Home

◆ Remember always that climate change is serious. How you respond is important. Smaller degrees of change in the past have destroyed civilizations. Every action is important. Reflect on your carbon footprint and seek to reduce it in every way possible.

◆ Become energy independent. Find ways to use solar or wind power instead of electricity from the utility grid.

◆ Use green energy. Green energy is environmentally friendly electricity that is generated from renewable energy sources such as wind and the sun. You can either buy green power from your utility or modify your house to generate your own energy - by installing a solar hot water heater or photovoltaic panels.

◆ Eat lower on the food chain. Whenever possible, buy locally grown, organic foods. Seek especially to avoid beef as it has a large impact on climate change.

◆ Travel less. Avoid unnecessary trips and refrain from unneeded long distance travel. Plan vacations close to home. Work close to home and avoid long commutes. Walk whenever you can. Transportation is one of the biggest causes of excess carbon dioxide in the atmosphere.

◆ Change your lights. Replace conventional incandescent bulbs with compact florescent bulbs or energy efficient LED bulbs. This single action will help you save on electrical bills.

◆ Heat and cool smartly. Clean air filters regularly. Lower the thermostat in winter and raise it in summer. When it's time to replace old equipment, choose a high efficiency model. Turn your water heater thermostat down to 120 F.

◆ Reduce, Reuse, Recycle, Repair and Refuse. Support the recycling programs in your community. Recycle newspapers, bottles, paper and other goods. Use products that can be recycled and items that can be repaired or reused. Buy products made from recycled materials. Reducing, reusing, and recycling helps conserve energy and restrains pollution and greenhouse gases.

◆ Unplug energy bandits. Unplug cell phone chargers, toasters and appliances when not in use. Turn off lights and other appliances when they are not in use.

◆ Use water efficiently. Municipal water systems use a lot of energy to purify and distribute water. Saving water, especially hot water, lowers greenhouse gas emissions.

◆ Study the data of climate change. Focus on sound, peer-reviewed science. A lot of junk science fills the media. For this reason, be careful to check your sources.

◆ Spread the word. Tell family and friends that energy efficiency is good for their homes and the environment because it lowers greenhouse gas emissions.

◆ Vote for candidates who understand the science, who appreciate the urgency of the challenge, and who will support responsible measures to address climate change.

At Your House of Worship

◆ Teach by example. Set a good example of care for God's creation by setting up a recycling program and by purchasing only recycled materials. This helps reduce emissions. Discuss climate-friendly choices as moral responsibilities.

◆ Purchase with awareness. Avoid throwaway products or anything that contains toxic substances. When possible, buy locally produced products.

◆ Conserve materials. Use both sides of paper. Cultivate a frugal attitude that respects the earth and its materials.

◆ Manage office equipment energy. Office equipment often uses energy even when idle or on stand-by. Always activate the power management features on your computer and monitor. Turn off equipment and lights when not in use.

◆ Invite high school students to calculate your house of worship's climate impact. Students can investigate the link between church activity, greenhouse gas emissions, and climate change. They can discuss ways to reduce the climate impact.

On the Road

◆ Travel less. Minimize optional or unnecessary travel. Drive a fuel efficient vehicle. Choose a vehicle that gets the best gas mileage.

◆ Give your car a break. Use public transportation, carpool, walk or bike whenever possible to avoid using your car. When possible, combine errands into one trip.

◆ Drive smart. Emphasize fuel economy; go easy on the brakes and gas pedal, avoid hard accelerations, reduce time spent idling.

> *Measures must be taken by each of us to reduce our impact on the world's climate. At minimum, this means caring about the effect of our lives upon our neighbors, respecting the natural environment, and demonstrating a willingness to live within the means of our planet.*

Mitigation and Adaption: The Long Term View

Climate change is forcing human society to rethink the economy and the policies that have lead to this global predicament. Climate change represents a failed economic vision that has created a global pathology of "climate disease" due to disregard of the ancient command to steward and replenish the Earth. Ten key concepts that can lead to a renewed societal vision and the regeneration of our democratic system of government include the following:

◆ Change lifestyles

Avoiding the worst impacts of climate change will require a fundamental shift in the way we generate energy and consume manufactured products. This policy shift should begin immediately and be well underway within the next ten years. This will mean a fee on carbon with perhaps a rebate to citizens. Community values will have to replace individualism, and moral and spiritual values will have to become priorities over crass materialism.

◆ Prioritize renewable energy

Government should subsidize the installation of clean, alternative sources of energy, especially solar and wind, wave, geothermal and other clean energy sources. Renewable energy technology can provide a far larger percentage of our total primary energy demand. Churches, temples, businesses and individuals, as a matter of right moral and ethical vision, should make the necessary sacrifices and switch over to renewable sources of energy. This should be part of the advocacy of individuals for use in municipalities and communities as well as recommendations to utilities and all houses of worship. Renewable energy is the morally correct energy and it is available now.

◆ Transform the vision

The scale of the climate challenge requires a transformation of the way we produce, consume and distribute energy. Fortunately, we have the technology to meet this challenge while giving a boost to the economy, energy sector employment and energy security. People must learn to cooperate and clean local food systems must arise. Organic whole foods must become a national priority.

◆ End corporate capitalism

We must reexamine every aspect in our lifestyle and begin to shift away from a profit-driven, corporation-sponsored consumer lifestyle to one that is local and community based. Corporate driven consumerism is clearly not sustainable. Whether we like it or not, several degrees of warming already exist in the climate "pipeline." This will require an adaptation to new genres of weather, water, food, and housing as well as community and economic relationships.

◆ Adapt to rising sea levels

The polar ice caps are melting. This will increase sea levels several feet in the near term, but much more in a century. Climate refugees will increase. Sea level rise will displace millions of people in coastal cities, including Boston, New York, Washington, Wilmington, Charleston, Miami, New Orleans, Galveston, San Diego, San Francisco, and many other coastal locations.

◆ Confront climate denialism

All efforts to block climate regulations and policies that would kickstart the clean energy revolution must be confronted. These efforts are most often financed and orchestrated by those involved with fossil fuel interests and through a network of surrogates. These selfish individuals must be confronted with the lack of integrity in this position and shown the folly of their ways.

◆ Maintain the global commons

Disruptions to the balances in ecosystems will harm everything from oceanside communities to minke whales to coral reefs to polar bears and cause a decline in human ability to appreciate the original handiwork of the Creator. Whole forests will be lost, and hundreds of thousands of species will become extinct. In response, forests and clean water must be treasured and restored as they sequester carbon and provide for clean water, stable hillsides, and moderate and moist terrain. Trees and forests must be restored and wilderness areas preserved. All animal and plant species must be respected and honored for the unique niches that they occupy in the grand scheme of creation. Schools must teach a practical form of creation care and the need for each person to be aware of a primal responsibility to restore and regenerate the land and its creatures.

◆ Cooperate locally and globally

Every person has a responsibility to act for a stable climate and to maintain the integrity of creation. Every community must be aware of the responsibility to serve immediate local issues as well as the global need to moderate CO_2 levels so that all may enjoy a liveable climate and the orderly progression of seasons and hospitable weather. A further responsibility exists for cooperation and sharing among people and nations. We share one life on one planet and all will inherit one future. Healing the climate requires a healing of relationships, from the local to the international level.

◆ Subsidize the Global South

Climate change will initially impact the global South more profoundly than those who live in the industrialized North. Because the climate crisis was created in the North, yet is wreaking havoc in the South, reparations are required to help those peoples most victimized by climate change to adapt to the harsher conditions which they are now already enduring.

◆ Restore a sacred vision of God within all things

President George Washington, in his Farewell Address (Philadelphia, 1796), emphasized the need of religion for the republic to survive. Here are his words: *"Of all the dispositions and habits which lead to political prosperity, religion and morality are indispensable supports. It is impossible to rightly govern the world without God and the Bible. In vain would any man claim the tribute of patriotism who should labor to subvert these great pillars of human happiness, these firmest props of the duties of men and citizens."*

> Washington is still right. Without religion, moral values decline and the integrity of government becomes subverted. Religion is distinct from "church." Separation of Church and state is one thing, as this implies a specific institution, but the generic practice of religion is essential for the maintenance of sound democratic institutions, as without virtuous citizens, we will only have corruption and malfeasance.

Cooperation and Networking

Working Across Religious, Social, Ethnic and Racial Differences

AS CLIMATE CHANGE INTENSIFIES, the harmony between society, land and weather will increasingly become disruptive and religious groups will find it useful to support each other and work cooperatively with each other and civic and social organizations. We are all in this climate predicament together and we will need each other so that our voices and actions can be amplified by working in groups and networks.

In the past it has often been difficult to work across religious traditions because of different theologies and a tendency for some groups to operate in isolation. However a lesson learned from hosting the annual National Prayer Breakfasts on Creation Care shows that through an interfaith protocol, every religious group can join with others and participate with integrity. Here are four principles that allow each faith entity to work in harmony and cooperation with others.

(1) INVITE each religious group to represent the fullness of the religious traditions which it represents. Ask each to do so without compromise in what they understand as their responsibility. With this understanding, each faith group can participate with the full strength and clarity of its approach to God and divinity. Ask each person to affirm if this is an acceptable request.

As a corollary to this protocol, (2) REQUEST that each person honor and respect every other participant's responsibility to represent the fullness of their faith tradition. This creates a safe framework for discussing the problems of climate change and how the changes that are now taking place equally affect us all, regardless of our religious differences.

Next, (3) ENCOURAGE each person to refrain from issues extraneous to climate change and creation care. This keeps the gathering focused on this crucial issue of mutual concern.

Finally, (4) ASK each person to avoid negativity and criticisms about one another, or their faith traditions, or issues in the past. With these four criteria in place, a foundation emerges that over time glues interreligious cooperation together.

With this formula the National Prayer Breakfasts on Creation Care and all of the Washington Week interfaith initiatives have grown and become a united and harmonious interfaith voice on environmental and creation care issues.

America's religious organizations hold remarkably similar perspectives on climate change and creation care issues generally. From National Forest policy to climate change, and from endangered species protection to the importance of thoughtful land management, the major religious institutions conclude amazingly similar policy perspectives. This allows the diversity of mainstream American religion, despite theological differences, to speak with one voice regarding public policy on climate and environmental issues. This is not political action so much as remedial moral and ethical education of our social and political leaders. The moral principles that shape faith traditions have always emphasized the necessity of clean air, clean water and clean food as fundamental requirements for healthy families and a healthy society. In this regard there is nothing new here.

What is new is the modern context. The intensity with which commercial interests press government to elevate their private good over and above the common good and the general welfare is in our day unprecedented. America's founding fathers saw religious institutions as an essential bulwark of a stable constitutional republic precisely because of its role in maintaining virtuous behavior and trust in God. Without this function in society, standards of personal behavior degrade into unbridled greed, selfishness and disregard for the welfare of the larger society.

In their examination of the role of religion in government, the founding fathers often declared the need for religion. They also discussed a separation of church from state. The implication is that the role of religion in society cannot be reduced to any single church nor should there be any alliance or special governmental relationship with any one faith institution.

The further implication is that the conflation of religion with church is a distortion of the historical assumptions about the foundations of good government. Mainstream religion has always held that any privatizing of the commonwealth (the air, water and essentials of life which all need) is wrong and contrary to healthy communities and good governance. Following this assessment of moral propriety and extending that to public policy, the National Religious Coalition on Creation Care has articulated an ethic of the environment which can guide individual actions, discern right from wrong in human impact upon the environment, and provide a deep critique of the actions of government or the individual.

This ethic of creation care encompasses and summarizes western religious traditions and allows for engagement with every environmental issue. In the spirit of the Scriptures this ethic is framed in the following terms:

> "The Earth is God's and all that is in it; thou shall
> not destroy the earth nor despoil the life thereon."

This ethic of creation is fundamental to Western religious belief and thought. It provides a means to critique every issue and allows perspective for shaping of personal lifestyles as well as public policy. When applied to environmental issues, it offers healing perspective and a vision for integrating human society into the ecosystems of the planet. In application it directs behavior in how we design society and commerce.

Over the past two decades, this ethic has allowed this interfaith coalition and its partners to further our religious vision on ecological problems. It has empowered those involved to bring the moral and spiritual vision of religion to government leaders; to conduct press conferences and congressional staff briefings, and generally to translate the understandings of religion onto discussion on national environmental policy. This same ethic can help you do the same.

A Key Action on Climate Change

Putting a Price on Carbon Based Fuels

By Donald Addu
Durham, North Carolina

THERE ARE MULTIPLE WAYS TO LOWER YOUR CARBON FOOTPRINT and fight global warming, including driving a more fuel efficient vehicle, installing LED lights, choosing local foods, and not investing in fossil fuel companies. Every time you purchase a product, you are voting for that product; you vote with your dollars. We can make the moral choice to reduce our carbon footprint, but if we make the moral choice and the economic choice the same, then we can truly beat our fossil fuel addiction.

The Citizens Climate Lobby promotes a harmony between the economic choice and the moral choice. A price on carbon-based fuels will help show the true cost of these energy products. We live in an artificial world, a world where the price we pay now hides the price we will pay in the future. We need a fee on carbon fuels to right this wrong. The carbon fee will be placed on forms of dirty energy, such as coal, oil and natural gas that are pulled out of the earth. The fee will also be assessed when that same energy is imported into the United States from across the world.

The best part of this plan is that the money collected from this fee will be returned to every American. This carbon fee and dividend will provide every family with a reimbursement check every month. This is your money; you can use to vote for a better tomorrow.

This "fee and dividend" strategy will make local food more affordable because it doesn't have to travel across the country. This fee and dividend will increase the demand for LED lights, dropping their cost and your electric bill. "Fee and dividend" will make our country safer by removing our dependence on foreign oil while adding 1.2 million jobs to our economy over 10 years.

We invite you to join our efforts and make a difference, not only by changing light bulbs, but also by reaching out to your community and faith leaders to educate them on the issue of climate change. We invite you to write a letter to the editor of your local newspaper about the need for climate action.

Most importantly, meet with your member of congress. They want to know how to help you and everyone they represent, but they won't know which plan to support if we don't tell them. We need both Republicans and Democrats to support climate action. Climate change impacts us all and it will take all of us working together to solve this problem.

A moral imperative, a citizen's climate bill, and a stable climate; all that is missing is you and your support. Thank you.

Notes:

RELIGION *and* GLOBAL CLIMATE CHANGE

A Handbook for Faith Leaders and Climate Activists

If you have found this book helpful, please send a copy to your pastor or someone who might benefit from this anthology. The price of a single copy is $12 plus $5 postage and handling. For 2 to 10 copies, the price is $10 each plus shipping of $5 for the first copy, and $2 for each additional copy sent to the same address. For copies sent to different addresses, please include $5 shipping for each copy. For 11 or more copies, the price is $9.00 each, plus appropriate postage.

A. Your name _____

 Address _____

 City, State and Zip _____

 Enclosed please find $_____ for _____ copies. I would like to include a donation to help promote the work of *Greening* Orthodox parishes. Optional donation_____.

                                       Total amount enclosed _____.

B. Please send _____ copies of **Religion and Global Climate Change** to the names below: (Use additional sheets of paper if necessary). Please include the purchase price of the book plus postage of $5 per book when mailing to different addresses. Thank you.

 Name_____   Name _____

 Address _____   Address_____

 City, State, Zip_____   City, State, Zip_____

 Name_____   Name _____

 Address _____   Address_____

 City, State, Zip_____   City, State, Zip_____

Publication Department

THE NATIONAL RELIGIOUS COALITION ON CREATION CARE

1100 Hughes Avenue

Santa Rosa, California 95407

Please make checks payable to: The NRCCC. Thank you.

www.ingramcontent.com/pod-product-compliance
Lightning Source LLC
Chambersburg PA
CBHW081258170526
45165CB00011B/3339